Sylvia J. Makowski

Sylvia J Makowski

37057 Mario Ann Court
Romulus, MI 48174-1274

W9-BPN-914

Managing Contract Quality Requirements

Also available from ASQ Quality Press:

Glossary and Tables for Statistical Quality Control, Fourth Edition
ASQ Statistics Division

Get It Right: A Guide to Strategic Quality Systems
Ken Imler

Principles of Quality Costs: Principles, Implementation, and Use, Third Edition
Jack Campanella, editor

Managing the Metrology System, Third Edition
C. Robert Pennella

Failure Mode and Effect Analysis: FMEA From Theory to Execution, Second Edition
D. H. Stamatis

Making Change Work: Practical Tools for Overcoming Human Resistance to Change
Brien Palmer

Business Performance through Lean Six Sigma: Linking the Knowledge Worker, the Twelve Pillars, and Baldrige
James T. Schutta

Leadership For Results: Removing Barriers to Success for People, Projects, and Processes
Tom Barker

The Supplier Management Handbook, Sixth Edition
James L. Bossert, editor, and the ASQ Customer-Supplier Division

Root Cause Analysis: Simplified Tools and Techniques, Second Edition
Bjørn Andersen and Tom Fagerhaug

Lean Kaizen: A Simplified Approach to Process Improvements
George Alukal and Anthony Manos

To request a complimentary catalog of ASQ Quality Press publications, call 800-248-1946, or visit our Web site at http://qualitypress.asq.org.

Managing Contract Quality Requirements

C. Robert Pennella

ASQ Quality Press
Milwaukee, Wisconsin

American Society for Quality, Quality Press, Milwaukee, WI 53203
© 2006 by ASQ
All rights reserved. Published 2006.
Printed in the United States of America.

12   11   10   09   08   07   06   5   4   3   2   1

Library of Congress Cataloging-in-Publication Data

Pennella, C. Robert
   Managing contract quality requirements / C. Robert Pennella.
      p. cm.
   Includes bibliographical references and index.
   ISBN-13: 978-0-87389-694-8
   1. Industrial procurement—Quality control. 2. Quality assurance—Standards.
   3. Contracts. I. Title.

   HD39.5.P46 2006
   658.7'23—dc22
                                        2006013168

ISBN-13: 978-0-87389-694-8
ISBN: 0-87389-694-7

Publisher: William A. Tony
Acquisitions Editor: Annemieke Hytinen
Project Editor: Paul O'Mara
Production Administrator: Randall Benson

ASQ Mission: The American Society for Quality advances individual,
organizational, and community excellence worldwide through learning,
quality improvement, and knowledge exchange.

Attention Bookstores, Wholesalers, Schools, and Corporations: ASQ Quality
Press books, videotapes, audiotapes, and software are available at quantity
discounts with bulk purchases for business, educational, or instructional use.
For information, please contact ASQ Quality Press at 800-248-1946, or write
to ASQ Quality Press, P.O. Box 3005, Milwaukee, WI 53201-3005.

To place orders or to request a free copy of the ASQ Quality Press Publications
Catalog, including ASQ membership information, call 800-248-1946. Visit our
Web site at www.asq.org or http://qualitypress.asq.org.

♾ Printed on acid-free paper

Quality Press
600 N. Plankinton Avenue
Milwaukee, Wisconsin 53203
Call toll free 800-248-1946
Fax 414-272-1734
www.asq.org
http://qualitypress.asq.org
http://standardsgroup.asq.org
E-mail: authors@asq.org

ASQ
AMERICAN SOCIETY
FOR QUALITY™

Contents

List of Figures

Preface

We believe that a look at the contract that joins customer and supplier is an opportunity to examine that relationship and to focus on the preparation and delivery of high-quality products and services. Typically, the focus for quality interventions rests on the supplier. We suggest that focus might well be expanded to look at the collaborative roles of both customer and supplier as agents for quality. In the chapters ahead, we would like to explore these roles as they contribute to a synergistic relationship where the sum of the parts exceeds the whole. The application of a purchase plan, communication of a purchase plan, application of a quality plan, verification of contract compliance, and the audit of contract compliance are five important elements of contract quality management that lead customer satisfaction.

Customer satisfaction is achieved when:

- Supplier and customer adopt collaborative strategies and common language
- Customer implements an objective purchase plan
- Supplier prepares an all-inclusive quality plan
- Contract is put in place that creates and supports common expectations as contract quality requirements are reviewed
- Customers and suppliers establish a business relationship that results in the production of high-quality products and services
- Both the customer and supplier win and are satisfied
- Customer is well informed about quality systems

- Customer has adequately prepared for a purchase
- Supplier has fully complied with specified contract quality requirements

The first five chapters in this book are listed sequentially in relation to the administrative application of contract requirements. (See Figure P-1.) The information contained in the this book is complete enough to be used as a seminar and teaching tool, and is designed to facilitate a mutual understanding of the needs, interests, and expectations of both buyer and seller.

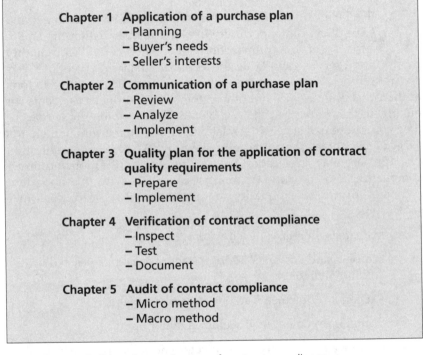

Chapter 1 Application of a purchase plan
  – Planning
  – Buyer's needs
  – Seller's interests

Chapter 2 Communication of a purchase plan
  – Review
  – Analyze
  – Implement

Chapter 3 Quality plan for the application of contract quality requirements
  – Prepare
  – Implement

Chapter 4 Verification of contract compliance
  – Inspect
  – Test
  – Document

Chapter 5 Audit of contract compliance
  – Micro method
  – Macro method

Figure P-1   Administrative application of contract compliance.

WHO WILL BENEFIT FROM THIS BOOK?

If you have to prepare contract quality requirements for a supplier or you are a supplier who will have to live with these requirements. If you are affected in any way by a contract or if you try not to think too much about contract quality requirements, then *this book is for you.*

Readers of this book will be able to:

- Establish and implement contract requirements effectively

- Identify and resolve actual and potential contractual problems

- Preclude overlapping of administrative efforts

- Reduce unanticipated costs associated with errors of omission

- Be better prepared for the administrative application and final outcome of contract quality requirements

BENEFITS TO PURCHASERS AND SUPPLIERS

Large and small organizations involved with requests, tenders, and contract administration want assurance that communication of requirements is clear, complete, and presented in a manner that eliminates errors of omission and misunderstanding. Organizations involved with outsourcing of goods and services are dealing with a marketing area that is best supported with the information referenced in this book.

- If you are a purchaser, you will benefit by learning to develop and implement objective documentation. Specifically, you will establish confidence that delivered supplies and services will conform to specified requirements. You will also be able to make objective decisions regarding the acceptability of products and services before a purchase is reduced to both cost and regret.

- If you are a supplier, you will find that the models for preparing and presenting your quality assurance capabilities will make you more prepared and competitive.

- Purchasers and suppliers involved with the application of quality system standards published by the American National Standards Institute, the International Organization for Standardization, the American Society for Quality, or equivalent quality system standards will be able to use the ideas in this book to support their efforts. All will benefit from the emphasis on pre-contract collaboration that ensures that quality proceeds from early expectations through delivery.

Our goal is to provide both purchasers and suppliers with a common language that will enhance their individual and common efforts by assuring that the supplies and services subject to quality assurance actions are adequately described and documented from the outset of the business relationship.

OTHER AUDIENCES

The book is also written for:

- Organizations that verify the capability of outside sources of supply through a first- and second-party arrangement

- Organizations preparing for ANSI/ISO/ASQ Q9001-2000: Quality Management Systems: Requirements or ISO 13485:2003 Quality Management Systems – Medical Devices, system requirements for regulatory purposes second- or third-party certified status

- Organizations subjected to initial or continuing quality audits

- Schools training students in the field of contract quality management

- Organizations involved in contract quality and quality systems intern programs

- Technical societies, such as the American Society for Quality, that are involved with education programs associated with the discipline of contract quality management

- Readers who wish to gain ready access to ideas and models that work

- Organizations involved in the application Total Quality Management programs

- Organizations seeking Six Sigma green- and black-belt certification

MAKING CONTRACTS WORK: PREPARATION, COMMUNICATION, AND COLLABORATION

Contracts protect and direct relationships. However, a true focus on quality goes beyond a contract to a spirit that is imbued within the organization by the leadership and contract administration on both sides of the negotiated agreement. While not losing sight of the importance of profitable relationships for both parties, management that views delivery of quality as a *two-way* street will generally be best prepared for success. The knowledgeable and proactive client is an asset to the supplier because that client knows what to ask for of a partner in the development of a project. Knowledgeable purchasers are less likely to fall back on the assumptions and vague recollections of promises associated with quality or, for that matter, any issue. The supplier who possesses an understanding of quality principles and who respects the importance of applying those sound principles will be able to define its responsibilities and limitations and more effectively describe the costs and benefits associated with these steps at the beginning of the relationship. In sum, when quality principles are explicitly understood and made part of an agreement, all stand to win.

Preparation by the customer and supplier also produces greater results. For example, when the customer prepares a clear technical data package and the supplier is able to read, understand, and implement, there will be less ambiguity and confusion. Key management personnel who are knowledgeable regarding the technical requirements of a contract will find the communication process between customer and supplier less complicated; key results are routinely accomplished correctly the first time, thus precluding costly and unnecessary repair or rework of nonconforming product and service characteristics.

A purchase agreement is the legal means to ensure that the purchaser and supplier have a common understanding of their rights and responsibilities. For the purpose of this book's presentation, the purchase agreement will be referred to as a *contract*. Whatever the terminology, issues fostering common understanding are essential to prevent misunderstanding at best and serious legal and financial problems at worst for either party. One of the key issues that requires a common understanding within a contract that effectively documents quality considerations, begin with clear and unambiguous communication that clusters around eight issues: management of policy; procedures and responsibilities; product design, development, control, and maintenance; marketing and servicing; administration of contract requirements; control of process; evidence of inspection and testing; and packaging, shipping, and billing.

Out of necessity, certain parts of the book may be construed to be speaking more directly to the supplier. However, the client who understands this principle will immediately increase his or her ability to meet the key result, that is, the delivery of products and services that meet or exceed expectations. When supplier and purchaser are well prepared and possess a common language for communicating about quality issues, communication and collaboration are more likely to be evident. The pages that follow are designed to facilitate common understanding.

Chapter 1, Application of a Purchase Plan—identifies the importance of establishing a purchase plan for each new product design. It also describes the physical characteristics that are required to meet customer and supplier needs, interests, and expectations. A purchaser establishes and coordinates a purchase plan for complex/critical items with support from a contract management team that is headed by a contract administrator. Support may be required from one or more specialists, and it is predicated on need. The team usually consists of specialists who monitor finance, production, purchasing, engineering, customer complaints, and safety.

Chapter 2, Communication of a Purchase Plan—outlines the review, analysis, and communication of a purchase plan. Communication begins well in advance of customer/supplier interaction. It starts with top management's written commitment to achieve nothing less than quality excellence with a competent work force that respects and develops everyone's capabilities. Commitment to quality must be supported with written policy and procedures that are workable, objective, and clearly understood. The extent of communicating contract requirements largely depends on the complexity of solicited products and services.

Chapter 3, Quality Plan for the Application of Contract Quality Requirements—describes the preparation and application of an objective quality plan that addresses the attainment of quality objectives and the financial rewards that are gained through the application of a robust quality plan.

Chapter 4, Verification of Contract Compliance—verifies compliance with contract quality requirements that is achieved by establishing, controlling, and maintaining the quality of products and services through all phases of a production cycle.

Verification of contract compliance associated with the acquisition of a product or service is greatly dependent on an organization's commitment to quality excellence. This is achieved with the preparation of a robust policy that describes an organization's written commitment to meet the intent of an adopted quality system standard and the preparation of complementary procedures and processes that are workable and objective. The checklists developed and used to verify the status of each contract line item become the documentation for the evaluation of contract requirements.

Chapter 5, Audit of Contract Compliance—describes how an organization audits its contract-related quality policy and procedures. This chapter addresses the merits of both macro and micro audits. It also provides guidance as to how and where they are implemented. Audits provide assurance to both internal and external customers that contract requirements are clearly defined and that established policy, procedures, and processes are, or are not, achieving defined objectives. The basis for auditing policy, procedures, and processes is the contract-related checklists that are prepared during a desk audit of a planned audit. It is at this point a determination is made that all policies, procedures, and processes have been considered, that they are in place, and that they are documented. Chapters 1 through 5 describe the needs and expectations of customer and supplier.

Chapter 6, Case Study 1: Processing an Invitation-for-Bid—describes *what* is required to satisfy the needs and expectations of both customer and supplier. It is important to note that a contract based on the lowest price alone can be a false economy; it can lead to unsatisfactory performance or late deliveries that result in added administrative costs to both customer and supplier.

Chapter 7, Case Study 2: Application of Contract Quality Requirements—addresses *how* needs and expectations are met. This is accomplished with narrative and graphic examples that augment the information contained in chapters 1 through 5.

Chapter 8, Self-Assessment of Managing Contract Quality Requirements Quiz—questions, answers, rational and source of rationale are provided to reinforce the information contained in this book and to assist readers who are seeking to improve their skills regarding the administrative application of contract quality requirements. This is accomplished with narrative and graphic examples that augment the information contained in chapters 1 through 5.

The questions *(Q)*, answers *(A)*, rationale *(R)*, and References *(Ref.)* are provided to reinforce the information contained in this book and assist readers who are seeking to improve their skills regarding the administrative application of contract quality management.

This book is a labor that reflects the help and support of many people. I owe an ongoing debt to Rosemary F. Garvey, the president of Blanchette Tool and Gage Manufacturing Company, for her willingness and ability to share her knowledge of the practicable application of quality management systems. She serves the profession well and has helped me greatly over the years.

I want to thank Donna Mugavero, vice president of Via Data Management and Services Corporation and a gifted problem solver, for her help, and Jeanne K. Derbyshire, regulatory manager, for sharing her knowledge and experience. I especially want to thank her for the time spent on editing this book.

To the Springfield Defense Contract Administration Management Office, a Division of the Defense Logistics Agency, I would like to express gratitude for the experience gained regarding the administrative application of contract quality requirements during my employment with them.

I am also grateful for the inspiration, sage critique, and guidance provided by Dr. Michael A. Pennella, a great son, who also edited the book and who was never too busy to take the time to provide support to me when I need it the most.

I acknowledge my colleagues and friends in the American Society for Quality, notably Edward Barabas, Frank Corcoran, Fred DeNude, and the members of the ASQ-North Jersey Section's executive board for their support and commitment to a strong profession.

1
Application of a Purchase Plan

1.1 CAPABILITY

A purchase is made from a buyer's list of capable suppliers. When there is a need to solicit a new supplier with unknown quality assurance capabilities, a proposed supplier must demonstrate to a potential client that it is capable of meeting specified requirements. Where applicable, the purchaser may also request that a proposed supplier verify its financial and production capabilities.

1.2 PURCHASE PLAN

A purchase plan is a precursor to soliciting capable suppliers of products and services. It is during the early stages of the planning process when a determination is made as to whether it is necessary to prepare a new plan or to modify one that was previously prepared. Plans are developed in conjunction with all other functions of contract management. A robust purchase plan is one that not only meets a purchaser's needs and expectations but also meets a supplier's interests. When developing a purchase plan, the planner receives support from specialists who are responsible for significant aspects of a purchase. Their support is particularly important where solicited supplies and services are complex.

1.2.1 Complexity applies to both product- and service-related contract line items. For example: A large missile may be classified as complex, and a service contract to repair that missile may also be classified as complex. Conversely, an *O*-ring is classified as a noncomplex product. Services regarding the input of information into a computer may also be classified as noncomplex.

1.2.2 In an attempt to reduce contract administrative costs, particularly where an organization is involved in the purchase of items that are simple, noncomplex, and noncritical, contract administration

1

may be delegated to an individual such as a purchasing manager or to a key representative of an organization. When a requisition is made for off-the-shelf items or items that are non-complex, the planner should decide if the selection of a supplier will be determined by a low bidder's past quality history producing like or similar items rather than issuing an invitation-for-bid to the public.

1.2.3 A purchaser establishes and coordinates a purchase plan for complex or critical items with support from a contract management team (CMT) that is headed by a contract administrator. Support may be required from one or more specialists and it is predicated on need. The team usually consists of specialists who monitor finance, production, purchasing, quality assurance, customer complaints, and safety. (See Chapter 5, Figure 5.8.)

1.3 CONTRACT ADMINISTRATOR

With support from the team of specialists, a contact administrator is responsible for the administrative application of contract quality requirements (CQR). He or she is the point of contact regarding contract compliance.

1.4 MONITOR OF CUSTOMER COMPLAINTS

A monitor of customer complaints maintains a running account of justified complaints to assure that causes of reported nonconformance are identified and rectified.

1.5 DIRECTOR OF FINANCE

The director of finance determines whether a new supplier of products and services has the required financial resources to support a proposed contract. If not, the task is to determine whether the supplier has made arrangements with an outside source to satisfy financial obligations. If a supplier is found to be financially capable, a positive recommendation for a contract award will made to the purchasing manager. Conversely, a supplier may be *technically* capable of producing a product or service but *may not* have the required financial resources or may be unable to obtain them from an outside lending agency. Experience dictates that under this condition it will be in the best interest of the proposed supplier and the potential customer for the director of finance to submit a negative recommendation for a contract award.

1.5.1 Where there is a need to verify financial capability, a request should be included in the solicitation for products and services. This is particularly important where a proposed supplier might be reluctant to release this information.

1.6 PRODUCTION SPECIALISTS

Production specialists determine the adequacy of plant facilities, production equipment, and delivery schedules. Their focus is centered on:

- Total manufacturing space

- The availability of storage space for production lots

- Lots accepted/rejected

- Maintenance of good housekeeping

- Availability of sufficient power and fuel to adequately meet production requirements

- Availability of an alternate power and fuel source where appropriate

- Availability of adequate material handling equipment

- Availability of transportation resources required for shipping the product to specified destinations

1.6.1 The production department provides assurance that human resources needs are also addressed. This includes determining the number of skilled, unskilled, engineering, and administrative personnel required to produce a product; identifying whether new personnel may be needed to produce the product; and, where appropriate, determining the shifts on which the work is to be performed. They also review methods by which a supplier's production plan and production resources will blend to assure delivery of products and services within the time constraints of a contract.

1.7 PURCHASING MANAGER

1.7.1 A purchasing manager, with help from a staff of specialists, reviews purchase requisitions to assure that:

- Technical requirements, quality system standards, specifications for inspection, testing, and other contract requirements essential for verifying the integrity of supplies and services are properly identified

- Tailored specifications are properly defined
- The need for audits is identified
- Capability of the selected supplier is verified

1.8 DIRECTOR OF QUALITY ASSURANCE

With support from a team of specialists, the director of quality assurance verifies that a proposed supplier is capable of meeting specified quality system standards. The standards include, but are not limited to:

1.8.1 ANSI/ISO/ASQ Q9001-2000 Quality Management Systems: Requirements. This standard specifies requirements for a quality management system that can be used to address customer satisfaction by meeting customer and applicable regulatory requirements.

1.8.2 ISO 10012-1: 1992, Quality Assurance Requirements for Measuring Equipment – Part I: Metrological Confirmation System for Measuring Equipment. This standard specifies the main features of a calibration system.

1.8.3 ISO 13485-2003: Quality Management Systems – Medical Devices: System Requirements for Regulatory Purposes. This standard provides a focus for the quality management system requirements for medical devices.

1.8.4 ANSI/ISO 14001-1996: Environmental Management Systems – Specifications with Guidance for Use.

1.8.5 ISO/TS 16949-2002: Quality Systems – Automotive Suppliers. Particular requirements for the application of ANSI/ISO/ASQ Q9001:2000.

1.8.6 ANSI/ISO 17025-1999: General requirements for the competence of testing and calibration laboratories. This standard specifies requirements for the competence to carry out tests and/or calibrations.

1.9 SAFETY SPECIALIST

Safety specialists are included as members of a contract quality management team when there is a need to verify the adequacy of safety requirements associated with hazardous or toxic items, radiation materials, and biological agents.

1.10 LEVELS OF PURCHASE

1.10.1 There are two levels of purchase associated with a product or service. Both must be considered when preparing a purchase plan:

1) *Complex items [that] have quality characteristics not wholly visible in the end item, for which contractual conformance must be established progressively through precise measurements, tests, and controls applied during purchasing, manufacturing, performance, assembly, and fabrication as an individual item or in conjunction with other items.[1]*

2) *Noncomplex items [that] have quality characteristics for which measurement and test of the end item are sufficient to determine conformance to contract requirements[2]*

1.11 PREPARING A PURCHASE PLAN

When preparing a purchase plan for a new product or service, the planner should take into account items previously produced that are similar in nature to determine if:

- There is a need to prepare new procedures or processes
- Previously produced procedures and processes are acceptable
- Procedures and processes previously prepared require a change or modification

1.11.1 A purchase plan is prepared shortly after the distribution of a timely and all-inclusive purchase requisition. When an acquisition is made for an item that is similar to one previously produced, the purchaser, with support from appropriate members of a CMT, will determine if existing procedures require any changes or modifications. Whenever significant changes are necessary, the planner should review the plan and make appropriate revisions. When changes are called for, there should be assurance that they will meet a requesting activities' needs and expectations, that they reflect a realistic performance schedule, and that they recognize the urgency of need.

1.12 REQUESTS, TENDERS, AND CONTRACTS

1.12.1. A strong purchase plan is one that includes procedures for the review of requests (invitations-for-bid), tenders (responses to an invitation-for-bid), and contracts (contract administration).

1.13 REQUESTS

Several factors should be addressed when preparing procedures associated with processing an invitation-for-bid for the purchase of products and services. Consider the following questions:

- Is there a clear description of solicited products and services?

- Are required drawings, specifications, and standards available when needed?

- Is the method of packaging, marking, and shipment of products identified?

- Are inspection and acceptance points of products and services identified?

- Will there be a need for a post-award orientation conference?

- Will there be a need for the client to conduct quality audits at a supplier's facility?

- Is the client given freedom regarding the investigation of nonconforming contract quality requirements at the supplier's facility?

- Where applicable, are supplementary contract quality requirements identified?

Note: The purchaser (first party) is responsible for verifying second-party capability and the second party is responsible for determining subcontractor (third-party) capabilities.

1.13.1 It is a buyer's prime responsibility to understand and communicate clearly what it expects from a supplier. This responsibility begins with the development of an all-inclusive purchase requisition. (See Figures 1.1 and 1.2.) When a purchase requisition is properly developed, a buyer can confidently select capable suppliers of solicited products and services. The supplier, in turn, will be able to perform its contractual obligations by tendering to its client, in a timely manner, products and services that conform to contract requirements. When a supplier fully understands the needs and expectations of its client, it will be better prepared to provide that organization with an indication that solicited products and/or services have met specified requirements.

1. Requested by_____

2. Line items: a. product_____ _____

                    b. service_____ _____

3. Requisition date_____ 4. Job no. _____ 5. Revision date_____

6. Recommended vendor(s)_____

_____

7. Delivery: to storage_____ other_____

8. Source inspection: ☐ yes   ☐ no

9. Certified inspection/test data: ☐ yes   ☐ no

10. Item no._____ 11. Quantity_____ 12. Item description_____

13. Technical documents:
    Title   Issue date

_____

_____

14. Preferred delivery date_____ 15. Promised delivery date_____

16. Allocation: Job no._____Quantity_____Contract no.(s)_____

17. Terms_____18. F.O.B. point: Source_____ Destination_____

19. Buyer_____ 20. Quantity purchased_____ 21. Price_____

22. Remarks_____

_____

23. Approved: Name_____Title_____Date_____

Figure 1.1   Purchase requisition.

Block number	Action
1. Requested by	Enter person's name and department.
2. Line items	a. Enter requested product(s). b. Enter requested service.
3. Requisition date	Enter as appropriate.
4. Job number	Enter as appropriate.
5. Revision date	Enter the date that the requisition was changed and the nature of the change in the remarks block.
6. Recommended vendor(s)	Enter only where vendor capability is known. When vendor capability is unknown, this information shall be obtained by the director of purchasing in cooperation with the director of quality assurance along with other applicable specialists and noted in Block 22.
7. Delivery	a. Enter a check mark in *to storage*, where applicable. b. If there is an immediate need for the product, route it directly to the production line and enter the destination.
8. Source inspection	Enter applicable check mark where inspection will be performed (at buyer's or supplier's facility).
9. Certified inspection and test data	Enter check mark as appropriate. Enter specific requirement in the remarks block.
10. Item number	Enter the item number referenced in the prime contract.
11. Quantity	Enter number of products requested. If a request is for service enter *See Remarks* and explain in Block 22.
12. Item description	Enter a brief description of the item. If additional space is needed, use *Remarks* block. If description is lengthy, use General Purpose Continuation Sheet. (See Figure 3.8.)

Figure 1.2   Instructions for completing purchase requisition.   *(Continued)*

(Continued)

Block number	Action
13. Technical documents	List applicable drawings, specifications, and standards and the issue date for each document. If more space is needed, continue on General Purpose Continuation Sheet.
14. Preferred delivery date	Enter date that will be in consonance with general production requirements and/or the delivery date specified in the contract.
15. Promised delivery date	Enter the delivery date agreed upon between the purchasing department and the supplier.
16. Allocation	Enter the job number, extra quantities, applicable contract number to where the product will be allocated. If the quantity is more than the quantities required for a particular contract line item and there is an anticipated need for identical items for future use, then so note the destination of extra quantities in Block 22.
17. Terms	Enter discount terms agreed upon between purchaser and supplier.
18. F.O.B. Point	a. Enter a check mark in the *Source* block, if F.O.B. *(free on board)* is to the supplier's loading dock. b. Enter a check mark in the *Destination* block if F.O.B. is to the customer's loading dock.
19. Buyer	Enter contact name, telephone, and e-mail.
20. Quantity purchased	Enter as appropriate.
21. Price	Enter price agreed upon between the buyer and supplier.
22. Remarks	Enter supplementary contract or any other special requirements.
23. Approved	Enter name and title of the director of purchasing or his/her delegated representative and date of approval.

Figure 1.2   Instructions for completing purchase requisition.

1.14 TENDERS

Objective quality systems management is based on the receipt of tenders from suppliers of products and services, who proactively respond to the following questions:

- Does the proposed supplier have adequate quality assurance, production, and financial resources to meet solicited requirements or the ability to obtain them from an outside source in a timely manner?

- Does the proposed supplier have a satisfactory performance record?

- Does the proposed supplier have the required organization, experience and technical skills, the required production, technical equipment, facilities, or the ability to obtain them?

- Where a promise is made by a supplier to acquire resources from an outside source, are they identified and backed up with written quotes?

- Does a submitted tender offer clearly address all of the information contained in an invitation-for-bid?

- Are all responsive conditions of a solicitation identified?

- When a solicitation for a product or service requires a supplier to post a performance bond, is there an affirmative response to this requirement?

- Are solicitations distributed to bidders with known quality assurance capabilities? If not, are arrangements made to permit the client's authorized representative to conduct a pre-award audit to verify the supplier's capabilities?

- Are responses to solicitations reviewed by a client's standing review board?

- Where applicable, are provisions made for the performance of a post-award orientation conference between client- and supplier-authorized representatives?

- Does the supplier have a satisfactory record of integrity and business ethics?

1.14.1 Note: When required resources are not available at the time of a pre-contract audit, a proposed supplier can meet solicited requirements by acquiring them from an outside source. However, delivery of promised resources must be provided with written quotes that are readily available and considered acceptable by an auditor.

1.14.2 Bidders are required to:

- Review drawings, specifications, standards, and other specified requirements of a statement of work.

- Show the unit price for each item offered.

- State definite time of delivery of supplies and services.

- Sign bid by an authorized representative of its organization.

1.14.3 A pre-contract audit is normally required when information is not readily available to the client to make a determination regarding the acceptability of a low bidder's tender offer. However, if a contemplated contract is for a small dollar value or it involves an off-the-shelf commercial item, there may not be a need to conduct an audit. Promptly after an award of a contract resulting from a tender offer, the client should notify the low bidder as well as unsuccessful bidders of its conclusion. This shall be accomplished via a written notice.

1.14.4 Bidders may e-mail bid responses. However they must be submitted within the time constraints of an invitation-for-bid and include quantities, unit price, time and delivery, and all other appropriate information required by the bid. In addition the proposed supplier must follow up the e-mail with an original signed copy of the bid.

1.15 CONTRACT ADMINISTRATION

Several questions are addressed by a supplier of products and services during the administrative application of contract quality requirements. For example:

- Are contract review procedures prepared and adequately implemented?

- Are contract review procedures coordinated with a contract quality management team (CMT)?

- Are contract quality requirements (CQR) reviewed and summarized?

- Are contract review actions documented?

- Are CQR distributed in a timely manner?

- Are required drawings, specifications, and standards available when needed for planning purposes?

- Does the CMT have a clear understanding of CQR?

- Are procedures and responsibilities established to rectify ambiguity, misunderstanding, and deficient contract requirements?

- Where there is a need to rectify a deficient contract requirement, is timely and satisfactory corrective action taken?

- When CQR are delegated to a subcontractor, are they documented in a satisfactory manner?

1.15.1 Note: Small organizations cannot afford the luxury of a full staff of CMT. Consequently, the role of contract administration is delegated to one or two individuals who *wear more than one hat.*

1.16 FLOW DOWN OF CQR

This must be documented and clearly defined by the purchaser and understood by suppliers of supplies and services. Distribution of CQR to applicable managers is accomplished only after it is determined that the associated contract and technical package is clearly defined. Each contract or purchase order should be reviewed and summarized to assure the scope of the requirements and related technical documents are understood by all management personnel who support the quality effort. Subsequent to the preparation of an abstract of CQR or a highlighted copy of a contract, recipients of the aforementioned documents and associated technical data package should identify requirements that require clarification. When it is determined that stated requirements are vague, ambiguous, or not clearly defined, an immediate post-award orientation conference (POC) should be convened between customer- and supplier-authorized representatives to rectify this condition. Where appropriate, invitation to participate in the POC should be extended to third parties by the supplier.

1.17 RECORDS

Records related to the administrative application of contract requirements include, but are not limited to, the following items:

- Approval/disapproval of waivers
- Billing

- Corrective actions

- Contract change notices

- Correspondence

- Customer complaints

- Investigation of customer complaints

- Milestone charts

- Orientation conference

- Product and quality system audits

- Record of shipments

1.18 BUYER AND SELLER

When complex items are purchased, collaboration among key members of a contract quality management team is important. Members of a buyer's CQT provide administrative assistance to its organization's contract administrator as to *what* contract requirements will be expected from a supplier. Conversely, each member of a supplier's CMT provides administrative assistance to its organization's contract administrator as to *how* each contract requirement will be met.

1.18.1 No purchase or contract award should be made until a buyer makes a determination of supplier responsibility. When exhibiting responsibility a seller must: indicate adequate financial resources; comply with a proposed delivery schedule; have a satisfactory performance record; have the necessary production, technical equipment and facilities (or the ability to obtain them); and have the necessary organization, experience and technical skills to meet contractual requirements. If it is determined that there is a need to look at the supplier's total administrative and manufacturing system, it should be so stated in a solicitation for a product or service; this precludes a low bidder's reluctance to do so after the receipt of a contract award. However, this should not be of concern when a product is purchased from an organization whose product is of a proprietary nature or to an organization that has previously achieved certified status from a first and second or from a first and third party.

1.18.2 Where appropriate, the customer's specialists in the various disciplines of contract management will evaluate the supplier's plans for meeting solicited requirements on behalf of its organization before finalizing a contract negotiation. They assure that policy and procedures are in place, that they are acceptable, that detail requirements contained in a statement of work and other parts of contractual agreements will be met, and that quality assurance requirements are properly applied.

1.19 CAPABILITY

A purchaser of products and services is cautioned not to completely rely on information presented in a potential supplier's brochure, without verifying statements referenced therein that is supported with objective evidence. Customers who are committed to objective quality management will team up with second and third parties that have achieved an acceptable quality history with them regarding the application of applicable quality system standards. (See list of standards, Chapter 1, Clause 1.8.)

1.20 PURCHASING SUPPLIES AND SERVICES

The customer must use good judgment when purchasing supplies and services. This is particularly important where contracts, drawings, and specifications are tailored to meet special requirements and when there's a need to check out a supplier's capabilities before making a commitment to purchase a product or service. A commitment is made with minimum risks when dealing with suppliers with *known* production, financial, and quality assurance capabilities. Also, financial benefits are achieved by a single-source supplier where there is a good quality history. When capabilities are *unknown*, discriminating customers will audit a proposed supplier's capabilities before a contract is awarded. If a low bidder (second party) with unknown capabilities declines to be audited by a first party, the second party might be considered nonresponsive to solicited requirements. If it is not feasible to audit a proposed supplier and this requirement is not a stipulated requirement of a buy, then a potential client must find other ways to determine competence and capability. That may mean relying on reputation and past experience. If the low bidder is the owner of a proprietary item there might not be a need to verify capability, particularly if the low bidder is an established producer of products offered to the public.

1.21 EVALUATING SUPPLIERS

The primary purpose of evaluating a supplier's competence and capability to perform on a contract is to assure the timely delivery of acceptable items. This action is taken with proposed suppliers where their financial, production, delivery, or quality assurance history with the potential customer is unknown. A purchasing manager, with support from the director of quality assurance and appropriate members of the contract quality management team, should verify the competency of each new supplier with questionable capability before a contract award is issued.

1.21.1 Prior the advent of ANSI/ISO/ASQ (ISO) first- and third-party certification and registration of supplier capability, the first- and second-party route was the only method of verifying supplier capability. It is recognized that the first- and third-party method is far less costly. However, some organizations are still content to operate under a first- and second-party agreement for the following three reasons:

- A purchaser is dealing with a supplier who has established an exceptionally good quality record within the industry.

- A purchaser is doing business with a supplier who is the owner of a proprietary item and has a history of delivering good quality.

- A prime contractor's quest for continuous improvement techniques requires direct contact with his or her suppliers.

1.22 TYPES OF SUPPLIERS

1.22.1 Producers of a product and organizations that provide services are listed in three general categories. The first list covers the name, address, and products offered by suppliers who have achieved ANSI/ISO/ASQ Q9001-2000 certified status. The second list identifies those suppliers who are *not* ISO certified, but have previously provided sufficient evidence of their capabilities in advance of a contract negotiation. The third category involves the acceptance of a product or service that is predicated on a certificate of conformance. A certificate of conformance may be used in certain instances at the discretion of a purchaser when the following conditions apply:

- Acceptance on the basis of a certificate of conformance is in the customer's best interest

- Small losses would occur in the event of a defect

- Because of the supplier's reputation or past experience, it is likely that the supplies or services furnished will be acceptable and any defective work would be replaced, corrected, or repaired without contest

- Supplier is an acceptable sole supplier

1.23 CONTRACT TYPES

Contracts are grouped into two categories: fixed-price contracts and cost-reimbursement contracts.

1.23.1 Fixed-price contracts provide for a price that is not subject to any adjustment on the basis of a supplier's cost experience in performing a contract. This contract type places full risks and responsibility for costs and profit or loss upon the supplier.

1.23.2 Cost-reimbursement contracts provide for payment of allowable costs, to the extent prescribed in a contract. They are suitable for use when uncertainties involved in contract performance do not permit costs to be estimated with sufficient accuracy to use a fixed-price contract.

1.24 SUPPLEMENTARY LIST OF SUPPLIERS

A supplementary list identifies new suppliers who have submitted a request to be included to a potential client's list of capable producers of products and services. The list is held in abeyance and addressed only when there is a need to seek additional procurement sources. The selection process is based on the following conditions:

- The supplier is ISO certified for products and/or services offered

- The supplier has a history of supplying quality products and services of the type solicited

- An audit of the supplier is required prior to adding its organization to a customer's list of approved suppliers of specific products and services

1.24.1 This list provides the customer with specific information about a supplier that includes:

- Date that the supplier was accepted as an approved supplier by an ANSI/ISO/ASQ Q9001-2000 Registrar

- Date the purchaser's quality assurance capabilities were approved under a first- and second-party agreement associated with products and services offered

- Name and address of the approved supplier

- Names of officers, owners, or partners

- The name of persons authorized to sign contracts and their official capacity, telephone, fax, and e-mail

- Identification of products or services offered by a supplier
- Number of years in business
- Manufacturing floor space in square feet
- Warehouse floor space in square feet
- Type of business (manufacturer, service establishment or regular dealer of off-the-shelf items)
- Date supplier accepted as a reliable supplier

1.25 SUPPLIER'S PERFORMANCE HISTORY

In addition to maintaining a list of capable suppliers, the purchaser—with support from the quality control department—should establish and maintain a system that can be used to evaluate a supplier's performance history. The application of that system can then be used to continuously evaluate each supplier's quality and delivery performance. Albeit a supplier's capability is normally evaluated prior to being accepted as a qualified supplier, quality-focused managers still need to be sure that established capabilities are continuously maintained and, where appropriate, continuously improved. This action can be triggered by the assessment of objective quality evidence furnished by the supplier and verified by the customer's quality assurance representative at the supplier's facility, at the customer's facility, or a combination of both. (See Figure 1.3, Supplier Performance Rating and Figure 1.4, Instructions for Preparing Supplier Performance Rating Form.) The primary purpose of an on-going evaluation of a supplier's performance is to provide feedback data to the director of purchasing regarding those suppliers who continually have an acceptable history as well as those suppliers who have an unacceptable history. One way to measure a supplier's quality performance is to establish a rating formula. This is achieved by dividing the number of acceptable production lots by the number of inspected lots. For example:

$$\frac{\text{Number of accepted lots}}{\text{Number of lots inspected}} \times 100 = \text{Rating percentage}$$

1.25.1 Supplier can then be rated as follows:

- A rating of 98 percent or better will be considered acceptable.

- A rating from 90 percent to 97 percent will be considered as acceptable only after satisfactory action is taken to correct the cause of observed nonconformances.

- A rating of less than 90 percent shall be considered unacceptable. (Consider seeking another source of supply.)

---

1. Type of organization: ☐ Manufacturer   ☐ Service organization

2. Name of organization _____

3. Address _____

4. Telephone _____ 5. Fax _____ 6. e-mail _____

7. Period of evaluation: From _____ To _____

8. Product/service identification _____

9. Date	10. Purchase order no.	11. Lot/unit	12. Accepted	13. Rejected	14. Rating
_____	_____	_____	_____	_____	_____
_____	_____	_____	_____	_____	_____
_____	_____	_____	_____	_____	_____
_____	_____	_____	_____	_____	_____

15. Comments _____

16. Distribution to:

   Director of purchasing _____ Date _____

   Executive management:
   Name _____ Title _____ Date _____

17. Prepared by: Name _____ Title _____ Date _____

Figure 1.3   Supplier performance rating.

Block number	Action
1. Type of organization	Enter as appropriate.
2. Name of organization	Enter as appropriate.
3. Address	Enter supplier's address and point of contact.
4. Telephone	Enter supplier's telephone number
5. Fax	Enter supplier's fax number.
6. e-mail	Enter supplier's e-mail address.
7. Period of evaluation	Enter period that a representative number of lots/items were checked.
8. Product/service identification	Enter a brief description of the product/service and technical documents (drawing/specification).
9. Date	Enter date production lot was inspected.
10. Purchase order no.	Enter as appropriate.
11. Lot/unit	Enter as appropriate.
12. Accepted	Enter as appropriate.
13. Rejected	Enter as appropriate.
14. Rating	Enter an accumulated rating for the period listed in Block 7 above by dividing the number of accepted lots by the number of lots inspected.
15. Comments	Enter appropriate comments.
16. Distribution	Enter as appropriate.
17. Prepared by	Enter the name and title of the director of quality assurance or designated representative.

Figure 1.4   Instructions for preparing supplier performance rating form.

1.26 CONCERNS OF CONTRACT ADMINISTRATORS

Customers and suppliers who are involved with prime and subcontract activity should have a clear understanding of the purchase plan in a manner that supports collaboration, communication, and superior quality. In order to create quality and synergy, customers and suppliers recognize they share common concerns. Therefore, contract administrators, with support from their team of advisors, need to address the following questions when preparing a purchase request:

- Are satisfactory procedures for quality assurance in place?

- How will records reflecting quality assurance actions be stored and accessed by the customer?

- Is a complete and up-to-date technical data package readily available?

- Has the customer's technical data package been screened for vague or ambiguous language?

- Are all technical documents included in the technical data package?

- How will the capability of subcontractors be verified?

- How will changes required in contracts, specifications, inspections, and related information be handled?

1.27 OTHER REQUIREMENTS

There are other important requirements that are included in a purchase plan. For example:

a) When raw material is purchased, the purchase plan should identify the applicable raw material specification. It should also include checks to ensure that each shipment of raw material is accompanied by a signed certificate of conformance signifying that the raw material meets specification requirements.

b) When it is determined that there will be a need for a supplier to furnish objective quality evidence in the form of an inspection/test report, this too should be stated in a contract or purchase order.

c) Where appropriate, the plan should include provisions for a purchaser's authorized quality assurance representative to audit its supplier's quality system as well as perform source inspection where appropriate.

1.28 MILESTONE CHARTING

A milestone chart that supports a documented purchase plan (see Figure 1.5) provides those who work within the process of purchasing new products and services with a better understanding of what is expected from them. At the same time, it puts them in an excellent position to fulfill their assigned responsibilities correctly the first time. In addition, they will be able to compare the documented process against the actual operation and make proactive recommendations to improve established procedures and processes. The purchase plan should also address milestones not completed during the application of the plan and reason for slippage. Actions taken regarding task slippage and recommendations made by those who work within a process should be assessed when establishing future purchase plans.

Topic	Schedule Completion Dates
1. Requisition of products and services	January 3, XXXX
2. Preparation of purchase plan	January 17, XXXX
3. Purchase plan approval	February 2, XXXX
4. Adequacy of technical data	February 16, XXXX
5. Issue invitation-for-bid	March 10, XXXX
6. Evaluate bid	April 7, XXXX
7. Verify supplier capability	April 14, XXXX
8. Identify qualified low bidder	April 21, XXXX
9. Audit of first low bidder	June 3, XXXX
10. Prepare audit report	June 7, XXXX
11. Attain contract signature	June 10, XXXX
12. Conduct post-award conference (as appropriate)	June 17, XXXX

Figure 1.5 Purchase plan major milestone schedule.

1.29 TWO NOTES OF WARNING

1.29.1 One: Where appropriate, a clause should be included in an invitation-for-bid that will permit the purchaser with an option to audit a potential supplier's quality assurance capabilities prior to a contractual commitment.

1.29.2 Two: Contingency plans must be in place to evaluate the capability of a second, and where appropriate, a third low bidder in the event it is determined that:

- The apparent low bidder in not an ANSI/ISO/ASQ Q9001-2000-certified supplier of solicited products and services

- It is determined via an on-site evaluation that the proposed supplier's capabilities are suspect and that timely corrective action cannot be taken

- The first low bidder is not familiar with the application of technical requirements and does not have a history of producing identical or similar items

1.30 FILES

Contract-related files should be considered closed when the last supplies and services have been delivered and accepted by the customer, and payments for products and services rendered are received by the supplier. Files shall be held in storage for a period of time determined by the supplier and/or the customer.

1.31 SUMMARY

A purchase plan tells employees of the purchasing department exactly what is expected from them. This includes quality system characteristics, product design, performance, and all other features of a statement of work. A clearly defined purchase plan that addresses all technical and nontechnical documents enhances the contract administration process. This, in turn, leads to the preparation of a quality assurance plan. Both are important processes associated with the administrative application of contract quality requirements. All of these actions are an imperative for meeting the needs, interests, and expectations of both customer and supplier.

NOTES

1. U.S. Department of Defense, General Services Administration, and National Aeronautics and Space Agency, *Federal Acquisition Regulation (FAR)*, part 46, clause 46.203 (b)(1), 1995.
2. *FAR*, part 46, clause 203 (b)(2), 1995.

2

Communication of a Purchase Plan

2.1 CONTRACT

"Contract means a mutually binding legal relationship obligating the seller to furnish the supplies or services and the buyer to pay for them. It includes all types of commitments that obligate the [customer] to an expenditure of appropriated funds and that, except as otherwise authorized, are in writing. In addition to bilateral instruments, contracts include (but are not limited to) awards and notices of awards, job orders, or task letters issued under basic ordering agreements, letter contracts, such as purchase orders, under which the contract becomes effective by written acceptance or performance, and bilateral contract modifications."[1]

2.2 COMMUNICATION

Communication begins well in advance of customer/supplier interaction. It starts with top management's written commitment to achieve nothing less than quality excellence with a competent work force that respects and develops everyone's capabilities. Top management's commitment to quality and customer satisfaction must be supported with policies and written procedures that are workable, objective, and clearly understood. The successful application of quality systems is based on clearly communicated contract quality requirements between an activity requesting a buy and a buyer; between a buyer and seller; between a seller and its team of contract administrators; and between the seller and its contract administrator and its subcontractors. (See Figure 2.1.)

2.3 COMPLEXITY

The extent of communicating contract quality requirements largely depends on the complexity of solicited products and services.

Complexity can range from end-item inspection and acceptance of non-complex items to a requirement for the implementation and continuous control of complex supplies and services. Therefore, in deference to cost effective and objective management, solicitation of articles from a manufacturer or an item from an organization that provides a service should be separated into the following three general categories:

Category	Complexity
1	Off-the-shelf items placed in stock by a distributor prior to receiving a contract
2	Noncomplex items, other than off-the-shelf items, with quality characteristics that are simple in nature
3	Complex articles that have quality characteristics not wholly visible in the end item, and service characteristics that are complex in nature

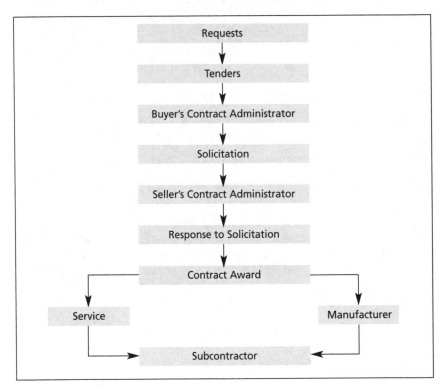

Figure 2.1   Communication of contract requirements.

2.4 PURPOSE

The purpose of grouping purchases of products and services is to identify those contract line items that will require special action, particularly those classified as Category 3. Required action for this category includes, but is not limited to:

- The need for a supplier to achieve certified status of personnel, equipment, or facility

- The need for concurrent first-article inspection and testing between seller and buyer

- The need to clarify certain contract quality requirements via an orientation conference between seller and buyer representatives

- The need to coordinate or resolve actual or potential problems

- The need to audit a seller's quality policy

- The need for buyers to inspect or test a product or services in advance of their tender offer for acceptance

- Potential losses associated with noncompliance with specified requirements

2.5 AUDIT

Contract content associated with the purchase of products and services is audited prior to releasing a solicitation to interested parties for bidding purposes. The primary purpose of an audit is to assure that technical and other pertinent requirements have been considered and that stated requirements are all-inclusive and free from omissions, ambiguity, or bias. An audit is also performed to assure that a prepared contract includes the detailed needs of the requesting activity in a most effective, economical, and timely manner. The customer's contract administrator and his/her technical advisors assume the responsibility for assuring that the needs and expectations documented in the purchase plan will be met. One method of assuring that expectations are not overlooked and that they are clearly communicated is to prepare a uniform contract format using, but not limited to, the topics listed in Figure 2.2.

Solicitation/contract number_____

Name and address of the purchaser_____

Business classification: Type of organization:
  ☐ Large corporation   ☐ Service
  ☐ Small business   ☐ Manufacturer
  ☐ Women owned   ☐ Off-the-shelf
  ☐ Minority   ☐ Dun's number
  ☐ Proprietary item

Issue date of the solicitation_____

Closing date of the solicitation_____

Number of pages_____

Space provided for the identification of bidder_____

_____

Address to where the product and/or service will be delivered _____

_____

Payment address_____

_____

Supplies and services: A brief description of supplies and services which includes the item number, quantities, unit price, extended costs, and where applicable, software and tailored requirements_____

_____

Standard documents. Enter a check mark where appropriate:
  ☐ ANSI/NCSL Z540-1-1994: Calibration laboratories and measuring and test equipment: General Requirements
  ☐ ANSI/ISO/ASQ Q9001-2000: Quality management systems: Requirements
  ☐ AS9100:2004: Quality management systems: Aeronautics: Requirements
  ☐ ISO 10012-1:2003: Quality assurance requirements for measuring equipment – Part 1: Metrological confirmation system for measuring equipment
  ☐ BS EN ISO 14971-2001: Medical devices – Application of risk management to medical devices
  ☐ ISO 13485:2003: Medical devices – Quality management systems: Requirements for regulatory purposes
  ☐ ISO/TS 16949:2002: Quality management systems: Particular requirements for the application of ISO 9001-2000 for automotive production and relevant service part organizations

Figure 2.2   Contract format.   *(Continued)*

(Continued)

Standard documents *(continued)*. Enter a check mark where appropriate:
- ☐ ANSI/ISO 17025-1999: General requirements for the competence of testing and calibration laboratories
- ☐ Pre-award survey
- ☐ Post-award orientation conference required
- ☐ First-article approval required
  - a. By the supplier _____
  - b. By the customer _____
  - c. Concurrent _____ (by customer and supplier representatives)

- ☐ Product inspection at source   ☐ Product inspection at destination
- ☐ Product acceptance at source   ☐ Product acceptance at destination
- ☐ Mandatory (concurrent) inspection by the customer and supplier

Configuration control   a. customer _____ b. supplier _____

Certified status or equivalent experience
- ☐ Certified calibration technician   ☐ Certified quality auditor
- ☐ Certified quality engineer   ☐ Certified quality
- ☐ Certified mechanical inspector

Certified quality technician
- ☐ Certified software quality engineer
- ☐ Certified Six Sigma Green Belt _____ Black Belt _____
- ☐ Qualified Product List (QPL)

Note: QPL is the process by which products are obtained from a manufacturer or a distributor and are examined and tested for compliance with specified requirements prior to the receipt of a contract and then included in a list of qualified suppliers.

Reliability, maintainability, and demonstration plan _____

_____

Design development _____

_____

Customer-supplied product(s)_____

_____

Customer-supplied inspection, measuring, and test equipment_____

_____

Responsibility for maintaining the accuracy of customer-furnished inspection, measuring, and test equipment.   a. supplier _____ b. customer _____

Risk analysis: Estimate and evaluate risks, control risks, and monitor the effectiveness of the control_____

_____

Figure 2.2   Contract format.

2.6 ABSTRACT OF CONTRACT REQUIREMENTS

A supplier of a product or service establishes and maintains an effective contract-review procedure; it is developed in conjunction with all other management functions that provide a summarized version of a contract to key managers. The purpose of this procedure is to demonstrate recognition of the quality aspects of a contract and an organized approach to achieving them. One method of providing key quality managers with contract content is accomplished through the preparation and distribution of an abstract of contract requirements. Another method is to highlight a copy of the procurement document for items that are noncritical and/or noncomplex. This informs top management and operations personnel that all specified contractual requirements have been identified. The preparation of an abstract of CQR along with the identification of related technical documents provides an indication that contract requirements were reviewed and identified, thus reducing errors of omission. Examples of abstracts of contract requirements and instructions for completing this form are shown in Figures 2.3 and 2.4, respectively.

1. Customer name and address _____

2. Telephone_____ 3. Fax_____ 4. e-mail_____

5. Contract no._____ 6. Award date_____

7. Amendment(s)_____ 8. Discount terms_____

9. Supplies and services

_____

_____

_____

_____

10. First article (FA): ☐ Required ☐ Not required   11. FA delivery date_____

12. Product inspection: Source _____ Destination _____

13. Product acceptance: Source _____ Destination _____

14. Packaging inspection: Source _____ Independent
                                                packaging plant _____

Figure 2.3 Abstract of contract quality requirements.   *(Continued)*

(Continued)

15. Item no.	16. Quantity	17. Destination	18. Delivery dates
_____	_____	_____	_____
_____	_____	_____	_____
_____	_____	_____	_____

19. FOB delivery point: Source _____ Destination _____

20. Technical data package: Specifications, standards, drawings, supplementary contract quality requirements, tailoring, etc.

_____

_____

_____

21. Supplementary contract quality requirements

_____

_____

_____

22. Comments

_____

_____

_____

_____

_____

_____

23. Prepared by	Title	Date

24. Reviewed by	Title	Date

25. Audited by	Title	Date

Figure 2.3   Abstract of contract quality requirements.

Block number	Action
1. Customer	Enter customer name, address, and point of contact.
2. Telephone	Enter telephone and extension number.
3. Fax	Enter fax number and designated point of contact.
4. e-mail	Enter e-mail address.
5. Contract no.	Enter number referenced in the contract award.
6. Award date	Enter date referenced in the contract.
7. Amendment(s)	Enter the identification number and a brief description of the amendment.
8. Discount terms	Enter number of days and discount percent offered.
9. Supplies and services	Enter description referenced in the contract.
10. First article (FA)	Enter an appropriate check mark in the *Required* or *Not Required* block.
11. FA delivery date	Enter date referenced in the contract.
12. Product inspection	Enter a check mark when inspection(s) are conducted at the supplier's plant. If it is to be conducted at the customer's facility enter a check mark in the *destination* block.
13. Product acceptance	Enter a check mark in the *source* block when acceptance inspection of a product or service is conducted at the supplier's facility. If it is to be conducted at the customer's plant enter a check mark in the *destination* block.
14. Packaging inspection	Enter a check mark in the *source* block when packaging will be performed at the producer's facility. If the packaging is to be performed at an independent plant, it shall be so stated in the contract. Enter a check in the *independent packaging plant* block.
15. Item no.	Enter the item number that is referenced in the contract.
16. Quantity	Enter quantity referenced in the contract.
17. Destination	Enter delivery destination(s) referenced in the contract.
18. Delivery dates	Enter dates referenced in the contract.

Figure 2.4 Instructions for completing an abstract of contract quality requirements. *(Continued)*

(Continued)

Block number	Action
19. FOB point	Enter delivery point specified in the contract. For example: a. *FOB source* (supplier's loading deck). b. *FOB destination* (customer's designated destination).
20. Technical data package	Enter specified quality system standard, calibration system standard, specification(s), drawing(s), or any other applicable documents referenced in the contract.
21. Supplementary requirements	Enter drawing, specifications, standards, or any other contractual requirements that were tailored to meet the purchaser's needs.
22. Comments	Enter as appropriate, such as the requirements for a pre-award survey, post-award orientation conference, special inspection requirement, or supplementary contract quality requirements, etc.
23., 24., 25. Prepared, Reviewed, Audited by	Enter as appropriate.

Figure 2.4   Instructions for completing an abstract of contract quality requirements.

2.7 ERRORS OF OMISSION

2.7.1 Contracts, technical documents, and established policies, procedures, and processes that are free from errors of omission will eliminate many stumbling blocks that originate from contractual and related documentation that are suspect. Contracts and technical documents that are up to date, clearly defined, and readily available will allow quality managers to successfully carry out their delegated responsibilities. In addition, contract administrators (or their designated representative) will be able to furnish the director of quality assurance and production departments with a robust abstract of contract requirements. This course of action will greatly assist management personnel in accomplishing their assigned tasks economically, objectively, and in accordance with the timely needs and expectations of internal and external customers.

2.7.2 What to Watch Out For

The root cause of errors of omission can be traceable to many elements of a quality system. Therefore, management involved in the administrative application of contract requirements is expected to have thoroughly reviewed contract requirements. Where ambiguity or differences of opinion prevail, immediate corrective action must be taken. Where appropriate, a post-award orientation conference should be held among first, second, and third party to rectify observed problems.

Errors of omission prevail when management's commitment to quality is not addressed; when quality plans are not established or maintained; or when contractual requirements are overlooked during a contract review process.

2.7.3 Errors of omission can (and usually do) lead to nonconforming processes, poor product quality, and unnecessary repair, rework, and deviations. They also prevail also when the sales department is too anxious to improve the bottom line with a *when in doubt, ship it out* mentality. Therefore an organization must take every precaution to ensure that errors of omission will *not* creep into the initial planning and during the transition of contractual requirements to management and operations personnel.

2.7.4 Potential sources of errors of omission include, but are not limited to, the following examples:

2.7.5 Contract Review

   a. A contract is inadequately reviewed and summarized

   b. A contract is not summarized

   c. Untimely action is taken to rectify a contract found to be deficient (see Chapter 5, Figure 5.5)

2.7.6 Management Commitment and Responsibilities

   • Top management is noncommittal regarding the application of policy procedures

   • Top management is not provided with feedback data associated with the rectification of costly customer complaints

   • Middle management is not given responsibility and authority to implement a workable quality system (see Chapter 5, Figure 5.3)

2.7.7 When top management fails to establish and maintain an organization of capable specialists or fails to provide key managers with specific responsibility, authority, and empowerment, it can lead to product and systemic problems that will be detected by an external customer. This method of operation is not only costly, it can also be detrimental to any positive advantages that a supplier might have within the industry.

2.7.8 Quality Management Systems – Requirements

- An organization fails to address or integrate all of the elements of an adopted quality system standard

- An organization's commitment to a specified quality system standard is defined in its written policies and detail procedures complement the application of policy statements (see Chapter 5, Figure 5.4)

- The following is a representative example of quality management systems standards adopted by producers of products and services:

  – ANSI/ISO/ASQ Q9000-2005: Quality management systems: Fundamentals and vocabulary

  – ANSI/ISO/ASQ Q9001-2000: Quality management systems: Requirements

  – ANSI/ISO/ASQ Q9004-2000: Quality management systems: Guidelines for performance improvements

Modified versions of the ANSI/ISO/ASQ Q9001-2000: Quality management systems: Requirements standards include the following:

- SAE Aerospace Standard–AS9100 Quality management systems: Aerospace: Requirements

- ISO 13485:2003 Medical Devices – Quality management systems - Requirements for regulatory purposes

- ISO/TS 16949:2002 Quality management systems: Particular requirements for the application of ISO 9001:2000 for automotive production and relevant service-part organization

The specified requirements of the ANSI/ISO/ASQ Q9001-2000 standard are not completely compatible with the needs and interest of the respective aerospace, medical, and automotive organizations. Hence, there was an obvious need for the publication of modified versions of the ISO 9001-2000 standards. Modifications to the quality management systems relate to specific requirements of the respective organization. Special quality management systems were published to demonstrate an applicable organization's ability to deliver aerospace, medical devices, automotive, and related services that consistently meet customer and regulatory requirements.

- British Standard BS EN ISO 14971:2001. *This International Standard specifies a procedure by which a manufacturer can identify the hazards associated with medical devices and their accessories, including in vitro diagnostic medical devices, estimate and evaluate the risks, control these risks, and monitor the effectiveness of the control. The requirements of this International Standard are applicable to all stages of the life cycle of a medical device. This International Standard does not apply to clinical judgements relating to the use of a medical device.*[2]

Policy and procedures are prepared for the acquisition and delivery of hazardous materials. The procedure identifies management personnel who are responsible for appraising its employees of all hazards to which they may be exposed; relative symptoms and required emergency treatment where appropriate; and proper conditions and procedure for safe use and exposure. (See Chapter 5, Figure 5.24.)

- Quality Program Requirements Mil-Q-9858A, 1981. This standard applies to all supplies (including equipment, sub-systems, and systems) or services when referenced in the item specification, contract, or purchase order.

- Inspection System Requirements, Mil-I-45208A,1981. This specification establishes requirements for contractors' inspection systems. These requirements pertain to the inspection and tests necessary to substantiate product conformance to drawings, specifications, and contract requirements.

Federal Acquisition Regulation (FAR) Part 42 prescribes general policies for performing contract administration functions and related audits. FAR Part 46 prescribes policy and procedures to ensure that supplies and services acquired under government contracts conform to the contract's quality and quantity requirements. Included are inspection acceptance, warranty, and other measures associated with quality requirements.

2.7.9 Planning

A supplier fails to establish a specific quality plan or has failed to consider the following aspects of a contractual agreement when conducting quality audits:

- Required resources including plant facilities, equipment, inspection, and test equipment and skills needed to fulfill contract quality requirements
- Identification of long lead items
- Product design, production process, installation, servicing, inspection, test procedures, and related documentation
- Need for new procedures and processes
- The need to update policy, procedures, processes, inspection, and testing techniques
- The need to develop or acquire new instrumentation
- The identification of product verification stations
- The identification, preparation, adequacy and retention of objective quality records

2.7.9.1 Interaction with the client

- The need to achieve certified status of personnel, equipment, and/or facility
- Cost of quality versus cost of no quality
- Analysis of risks associated with hazardous devices

2.7.10 Purchasing

Failure to clearly identify and verify capable sources of supply and/or failure to properly identify a statement of work and associated drawings, specifications, standards, and other pertinent requirements in a contractual document will lead to noncompliance with specified technical and other applicable contract requirements. (See Chapter 5, Figure 5.8.)

2.7.11 Facilities and Equipment

Errors of omission are prevalent when a supplier fails to develop required plant facilities and equipment or fails to demonstrate that they can be acquired from an outside source in sufficient time to meet a contractual obligation.

2.7.12 Document and Data Control

Errors of omission for this element can be attributed to missing or inadequate written procedures that define the process for issuing and retrieving administrative and technical documents. They prevail when required documents are not readily available at a product/service verification station when needed and the use of *canned* documents. These errors can also lead to multiple revisions to engineering drawings. (See Chapter 5, Figure 5.7.)

2.7.13 Inspection and Testing

Failure to identify pertinent inspection and test characteristics; failure to establish clear, concise, and all-inclusive inspection procedures; lack of acceptance criteria; or the failure to upgrade existing inspection procedures can lead to shipping defective products and services that are ultimately detected by the end user of an item. This will trigger customer complaints, or even worse, the customer may *not* register a complaint to the supplier and decide to simply seek another source of supply. (See Chapter 5, Figure 5.12.)

2.7.14 Inspection and Test Status

When procedures are not established for the satisfactory identification of the quality status (good of bad) of a product, a supplier runs the risk of commingling conforming products with nonconforming products. Conforming as well as nonconforming items should be continuously identified and separated during all phases of a production cycle. When this element of a quality system is omitted from a quality system or quality plan, reactive management takes over, costs escalate, and a tainted image usually follows. (See Chapter 5, Figure 5.14.)

2.7.15 Process Control

The establishment and implementation of a robust process control program is an imperative for measuring variability and producing a product that is right the first time; it also establishes a foundation for improving the processes that create products and services with the application of statistical process control techniques. Failure to do so can lead to duplication of work and added cost to produce an item. (See Chapter 5, Figure 5.11.)

2.7.16 Control of Nonconforming Product

Errors of omission are prevalent when:

- Associated policy and procedures are not established or available when needed

- Nonconforming items are not identified, segregated, or placed in a designated holding area
- Untrained personnel, inadequate product identification
- Nonconforming items are released from a designated holding area by unauthorized personnel
- Inadequate product identification labels, associated with the medical industry (see Chapter 5, Figure 5.15)

2.7.17 Corrective and Preventive Action

An organization will experience errors of omission when:
- Procedures are not established and maintained for the early detection and correction of nonconforming products
- Established procedures do not provide for the identification and correction of justified customer complaints
- Person responsible for monitoring customer complaints is not identified
- Corrective measures are not taken to preclude a recurrence of justified internal and external complaints
- The root cause of nonconformance is not identified
- Preventive action is ignored
- Continuous improvement of the processes that create a product is ignored
- Loss of improvement opportunities results in loss of business (see Chapter 5, Figure 5.16)

2.7.18 Control of Inspection, Measuring, and Test Equipment

Errors of omission occur when:
- The contract administrator or any other person designated with this responsibility fails to furnish quality assurance personnel with timely copies of the latest technical documents required for the selection of required measuring instruments
- Tailored technical documents are not furnished to responsible personnel in a timely manner
- Policy procedures are inadequate or not readily available
- Calibration procedures are missing or inadequately maintained
- The source of calibration procedures is not identified
- Accuracy requirements of working instruments and related measurement standards are unknown

- The accuracy of measuring instruments is not verified with higher-level instrument standards of known accuracy
- The establishment of calibration intervals is nonexistent
- The adjustment of calibration intervals is ignored
- Significant out-of-tolerance conditions are not identified
- Traceability of the accuracy of measuring instruments by an unbroken chain of events to a national or an international standard is nonexistent
- Accuracy ratios between product characteristics and general purpose measuring instruments are nonexistent (see Chapter 5, Figure 5.13

2.7.19 Control of Records

The establishment, maintenance, and control of quality records are a main source of objective quality evidence. If they are missing, inadequately established and maintained, or a quality plan fails to address them, it can create a negative impact on objective quality systems management. Without records, an auditor will have no choice but to conclude that an evaluated requirement never happened. (See Chapter 5, Figure 5.18.)

2.7.20 Customer-Supplied Product

Errors of omission can occur when:

- Provisions for the control of customer-supplied products are not established or implemented (see Chapter 5, Figure 5.9)
- The product is not examined for count and condition upon receipt
- The product is not guarded against damage during storage and handling
- Customer-supplied product in not inspected or tested prior to inclusion into a product

2.7.21 Handling, Storage, Preservation, Packing, Packaging, and Delivery

Suppliers who fail to establish and maintain satisfactory procedures and controls for this element of a quality system create a condition that can lead to items damaged while in transit or storage. Preservation and packaging methods, shipping containers, and delivery services that are suspect create a condition that can lead to the delivery of damaged products; justified customer complaints and added costs are bound to follow. (See Chapter 5, Figure 5.17.)

2.7.22 Quality Audits

Errors of omission will occur when an audit is influenced by bias. Here is an example: An audit is conducted by an owner of a process rather than by a qualified person who is *not* involved with the process scheduled for auditing. Or an auditor elects to use a generic audit checklist rather than one that is tailored from contract-related policy, procedures, and associated quality requirements. (See Chapter 5, Figure 5.19.)

2.7.23 Product Identification and Traceability

Procedures that fail to address the traceability of a measured product, and associated measuring and test equipment to an applicable contract/purchase order impede identification of the root causes of a reported complaint. Without proper controls, the investigation of internal and external complaints can prove to be cumbersome and time consuming. Inadequate or lost product identification can lead to services mix-ups and impede traceability to a service. (See Chapter 5, Figure 5.10.)

2.7.24 Training

A supplier should establish and maintain a training program that identifies the training needs of employees who perform functions of contract quality management. Proper training is important because products and services are produced correctly the first time because of knowledgeable employees. Failure to establish an objective training program will not be in consonance with the knowledge that is needed to continuously improve the processes that create the products. (See Chapter 5, Figure 5.20.)

Information regarding the application of a training program is contained in standard ANSI/ISO/ASQ Q10015-2001: Quality management: Guidelines for training.

2.8 SOLUTION TO REDUCING POTENTIAL SOURCE OF ERRORS OF OMISSION

a. Conduct an in-depth review and analysis of contract quality requirements and associated technical documents

b. Assure that there is a clear understanding of contract and technical requirements before establishing appropriate policy and procedures

c. Prepare robust quality and purchase plans

    d. Prepare a milestone of major subjects (by someone or a team of individuals thoroughly familiar with contract and technical requirements)

    e. Continuously upgrade and improve the processes that create products and services

2.9 CUSTOMER COMPLAINT MONITOR

Justified complaints should be reviewed and summarized by a representative of the quality department such as customer complaint monitor (CCM). This action is imperative for identifying those trends that adversely impact the processes that create products and services. That individual reviews reports prepared by purchasing, engineering, production, and quality control department representatives who are involved with the corrective action process and assures that the causes of nonconformity are properly identified and required corrective actions are implemented in a timely manner throughout appropriate areas of a supplier's organization. This includes product verification stations, work areas, procedures, statistical techniques, processes, tooling, and any other resources that impact product quality. This is achieved by monitoring trends and analyzing them through the use of charts and graphs. Where statistical process control (SPC) techniques are used to measure and analyze processes that reflect undesirable trends, the CCM may also be given the responsibility for monitoring corrective actions taken to bring a process into accepted limits. Communication between the CCM and concerned parties includes the need for immediate corrective action, required follow-up action, a corrective action plan, the adequacy of corrective action taken, and the verification that corrective action is taken in accordance with planned arrangements.

2.10 CORRECTIVE ACTION

The extent of corrective action is dependent upon the seriousness and frequency of nonconformance and its impact on an end item. The supplier is also obligated to correct product defects and procedural deficiencies prevalent in his or her quality system, eliminate the root cause of all deficiencies, and make appropriate changes and/or additions. In addition, organizations should periodically audit related policy and procedures to assure that recommended corrective action is maintained. Areas of the audit activity might include the purchaser's facility, a supplier's plant, or a sub-supplier's facility.

2.11 COMMUNICATION

Communication of contract quality requirements is extended throughout the application of quality system requirements. It should be directly related to open contracts and address applicable elements of an adopted quality system standard and related factors.

2.12 CONCLUSION

When customer and supplier quality assurance representatives agree that procedures, processes, and associated documentation are in place, a reinforced and proactive contractual relationship is established. Documentation includes quality objectives, job operation sheets, shop travelers, tote-bin tickets, inspection and test records, laboratory data analysis, engineering changes, and engineering approvals. Other substantiation documentation includes a quality system manual, quality, purchase, and production plans, engineering handbooks, standing operation procedures, and procedures manual.

NOTES

1. Department of Defense, General Services Administration, and National Aeronautics and Space Agency, Federal Acquisition Regulation, 1995 Part 2, Subpart 2.
2. BS EN ISO 14971:2001 Medical devices: Application of risk management to medical devices.

3

Quality Plan for the Application of Contract Quality Requirements

3.1 PRODUCT REALIZATION

The organization shall plan and develop the processes needed for product realization. Planning of product realization shall be consistent with the requirements of the other processes of the quality management system.[1]

3.2 PURPOSE

The purpose of developing a quality plan is to establish a commitment that takes into account relevant systemic elements and associated factors of an adopted quality system standard so that nothing in the administrative application of contract quality requirements will be overlooked. The depth in which a quality plan is developed depends on the complexity of a product design and other actions considered necessary at the time a plan is being developed. Quality plans are developed after the receipt of a contract and associated technical documents. When new contracts are received that contain a requirement that will necessitate a change in established procedures and/or processes, an original plan will be modified accordingly. Suppliers of products and services prepare and maintain a quality plan that describes how its organization will ensure compliance with a contractual obligation. The quality plan serves as a master plan and a control document. It covers service and product design requirements, and it is reviewed periodically for improvements and updated accordingly. The planner, with support from his or her team of specialists, also reviews previously prepared plans for a similar product design to determine if the information referenced therein might be incorporated into the new plan. When significant changes occur in a product design or associated contract quality requirements, the planner must immediately review the plan and make appropriate changes. Documents and specifications should be readily available to management and operations personnel so that a plan can be established, controlled, and maintained within the time constraints of a contractual obligation. Quality objectives and financial rewards are gained through the application and maintenance of a robust quality plan that is blended together with the administrative functions referenced in the following paragraphs. Each should be considered when establishing a quality plan.

3.3 PROCESSES

The plan addresses important process measuring points; product characteristic(s) that will be used to measure process variability; the identification and traceability of measured product characteristics; and the identification and traceability of pertinent measuring instruments and measurement standards that are used to check the process. In addition the plan identifies where and how specified *special processes* such as heat treating, magnetic particle inspection, and welding will be inspected and controlled. Last but not least, the plan includes a process flow chart that lists the sequence of operations and inspections as well as the job number, item identification, drawing and/or specification number, lot quantity, yield quantity, contract or purchase order number and the time required for each operation.

3.4 PLANT FACILITIES AND EQUIPMENT

When it is determined that new plant facilities or equipment will be needed to produce contract line items, the plan will indicate if they will be developed within the prime contractor's facility or if they will be acquired from an outside source.

When a potential supplier's plant facilities and equipment are unknown, the following factors must be addressed:

- Size of tract
- Square feet under roof
- Description of building: owned or leased
- Number of buildings
- Available space: manufacturing, inspection, storage, and shipping
- Miscellaneous: good housekeeping, power and fuel supply, alternate power and fuel source, material handling equipment, and transportation for shipping products
- Required production equipment for manufacturing, special tooling, and special testing

3.5 METROLOGY

A quality plan includes the adopted calibration system standard; the applicable procedures; and required inspection, measuring, and test equipment.

An excellent source of reference material associated with calibration standards can be found in the *referenced materials* section of ASQ Certified Calibration Technician catalogue. A representative sample is as follows:

- ANSI/ISO 17025-1999: General requirements for the competence of testing and calibration laboratories.

- ANSI/NCSL Z540-1-1994: Calibration laboratories and measuring and test equipment: General requirements.

- Richard Calhoun, *Calibration and Standards DC-40 GHz.* Louisville KY: SS&S, 1994.

- Ernest O Doebelin, *Measurement Systems: Application and Design,* 4th ed.; New York: Mcgraw-Hill, 1990.

- C. Robert Pennella, *Managing the Metrology System,* 3rd ed. Milwaukee, WI: ASQ Quality Press, 2004.

3.6. HUMAN RESOURCES

When it is determined that additional personnel will be required to fulfill a contractual obligation and they are not readily available, the number, type, and acquisition source should be identified in the quality plan.

When a first party audits second-party human resources, focus is centered on the number and skill of employees. Of primary concern are the required number of skilled and unskilled production employees as well as engineering, quality assurance, and administrative personnel that will be required to fulfill a contractual obligation. Verification regarding adequacy of third-party personnel is the responsibility of the second party. It is important to note that although verification of the adequacy of third-party capabilities is that of a second party, it is still the responsibility of a second party to provide objective evidence to the first party regarding third-party capability.

3.7 SPECIAL CONTRACT QUALITY REQUIREMENTS

Modification of a specification that is a component of an original product design can be classified as a special contract quality requirement. Concurrent inspection and testing of a critical product characteristic by customer and supplier quality assurance representatives is another example of a special contract requirement. When a change notice or a modification to a purchase is issued, its impact on a completed quality plan must be determined.

3.8 RESPONSIBILITIES

Organizational responsibilities vary. For example, a planner for a large organization may receive support from several members of a contract management team. Conversely, employees of small companies often wear many hats; individual specialists are not available to administer the various disciplines of contract management. Responsibilities associated with the administrative application of contract requirements are centered on requests, tenders, and contracts. When these elements are clearly defined responsibilities of management and operations personnel, large and small organizations will be able to apply their respective responsibilities in a timely and efficient manner.

3.9 PRODUCT DESIGN AND DEVELOPMENT

3.9.1 Engineering personnel study procurement documents, a statement of work, and other applicable documents to be certain that the mission and parameters of design is accomplished. They develop an initial design for feasibility and analysis purposes, for both the general design and for preliminary reliability analysis. A project engineer will work during this phase to ensure that a developed functional diagram represents the proposed configuration. The design plan is periodically reviewed and revised, as required, to provide a current model of the design configuration. Engineering personnel are responsible for all decisions regarding the redesigning of a proposed configuration. They determine the effects of design changes and other techniques utilized to improve the intrinsic reliability of a design. They also establish criteria for the determination of a system's success or failure in terms of specific inputs, outputs, and other parameters compatible with the requirements of a detailed specification.

3.9.2 Engineering specialists participate in the review of all observed failures during qualification and reliability demonstration testing, and they are responsible for incorporating necessary design changes to restore an item to a specified reliability level. They also determine the cause of failures such as design deficiency, fabrication error, material deficiency, or any other failures related to product development, testing, or fabrication. All failures related to design deficiency, or failures that are known to be repetitive in nature and for which an attributable cause has been determined, are corrected on all production units as stipulated by specification requirements.

3.9.3 Engineering develops Standard Operating Procedures, Standard Test Procedures and Standard Manufacturing Procedures, concurrently with the design development. These procedures identify the acceptance criteria for both fabricated and purchased components. They are prepared in sufficient detail to permit skilled as well as relatively unskilled personnel to perform that function. The procedures should be developed showing the sequence of operation, measurements to be recorded, allowable tolerance in the measured values, and the inspection and testing equipment that will be required to examine the product.

3.9.4 A design review is conducted by a group of technical specialists who are familiar with the essential aspects of a design, component parts, human resources, finance, quality assurance, production, maintainability, and reliability requirements. A project engineer along with engineering and quality assurance specialists participate in a formal design review. The nature and/or complexity of a proposed design will dictate what specialists will be present during this review. Reports, recommendations, and other applicable documentation must be readily available to all design review participants. Informal design reviews may be accomplished on all proposed designs not included in a formal review. Integration of functions must be maintained between methods engineering, production, and quality control departments so as to detect potential fabrication problems, maintainability problems, or any other problem areas prior to the release of engineering drawings and specifications for production. Periodic progress meetings are held to discuss problems and solutions. All recommendations and corrective action reports must be readily available during these meetings. The complexity of a design will dictate the types of technical specialists who should participate in these meetings. The design review team should extend an invitation to key department managers with the opportunity to supply objective comments and recommendations at these meetings.

3.9.5 The project engineer also participates in the review of nonconforming parts and components with other members during material review board actions to determine the effect that a reported nonconformance has on performance, safety, and interchangeability of product design parameters. He or she is also concerned with the adequacy of recommended actions taken to correct the cause of a reported deficiency as well as the proper disposition of nonconforming items.

3.9.6 Engineering, with the support from the quality control department maintains a system for the control of experimental, engineering, manufacturing, tooling, drawings, and specifications. The system is implemented to assure that only current drawings and changes are delivered to applicable areas of a production process. An indication that the design review has been accomplished and that it is in compliance with specified requirements will include the signatures of the project engineer and the director of quality assurance. (See Chapter 5, Figure 5.6.)

3.10 FIRST ARTICLE

First-article (FA) review is accomplished to provide a product that is satisfactory for its intended use and thus minimize the risks of producing nonconforming products and services at the outset of an initial production lot. First-article approval is also accomplished to demonstrate that manufacturing processes employed, workmanship standards utilized, and methods employed for the control of product quality are capable of producing an item that meets a specified requirement. The inspection and testing of a FA will further indicate whether the producer of an item correctly interpreted specified technical requirements. Another reason for conducting first-article inspections is to assess the adequacy of materials and processes. They may be performed by both supplier and customer or by the supplier alone. Organizations responsible for conducting FA inspection and testing are obligated to prepare a report of each inspection and a test that will show the technical documents that were used to check the first article, including applicable specifications, drawings, and standards. In addition, the report addresses other information such as the applicable contract or purchase order number, company's name and address, where the first-article action article inspection and testing was performed, and the test results in quantitative and qualitative terms. When first-article inspection and testing is performed by the customer, it is identified in a contract. It may be conducted at the customer's facility, at the supplier's facility, or at an approved commercial testing laboratory. When a customer (first party) is responsible for first-article inspection and testing, the supplier (second party) will submit a sample or samples to its customer that conform to specified technical requirements. Before releasing a first article to the customer for verification and acceptance, the supplier will assure that all characteristics that can be checked at its own facility, as well as those that were delegated to a subcontractor (third party), have been verified as conforming to contract quality requirements. See Figure 3.1 for an example of a form that is used to conduct a first-article inspection.

Observation Record

Name and address of client _____

_____

Item nomenclature _____ Part number _____ Revision _____

Specification number _____ Drawing number _____

Contract number _____ Purchase order number _____

Type of inspection _____ [Concurrent] _____

Participants _____ [Buyer and seller representatives]

Procedure _____

_____

Product characteristic code no.	Characteristic identification	Measuring device (MD)	MD identification number	MD code number	Observations
_____	_____	_____	_____	_____	_____
_____	_____	_____	_____	_____	_____
_____	_____	_____	_____	_____	_____

Disposition _____

Prepared by _____ [Representative of the supplier's quality control department]

Participants _____ [Representatives of buyer and seller]

Figure 3.1   Record of first-article inspection.

3.11 DOCUMENT CHANGE CONTROL

A document change control procedure is established and maintained for the control of documentation associated with the preparation of a product design as well as for operational functions. The procedure is designed to implement controls over both the method of implementing a change and the individual responsible for distributing a proposed change. Engineering, under the surveillance of the quality control department, maintains procedures for the control of experimental, engineering, manufacturing, tooling, drawings, inspection and test instructions, and other technical documents such as specifications and standards. The procedures also verify that documents and changes are readily available to management and operation personnel. In deference to maintaining total control of quality documents, requests must be furnished via a written request. (See Request for Technical Document, Figure 3.2.)

Date_____ Requested by_____

Document no. _____ Description _____

Reason for request:

_____

_____

_____

_____

Received by_____Title_____Date_____

Figure 3.2   Request for technical document.

3.11.1 There are three conditions that should be considered when implementing a document change control procedure:

1. When a contract requires a supplier to produce a product in accordance with a customer's design, all document changes require the approval of the customer's quality assurance representative

2. Where qualification testing to a customer's product design is a contractual obligation by a supplier, changes also require the approval of the customer's designated quality assurance representative

3. All contract-related changes observed by a supplier should be coordinated with his or her client

3.12 TECHNICAL DOCUMENTS

Drawings and specifications are prepared in sufficient detail to permit management and operations personnel to perform their assigned functions in a timely manner. Procedures regarding the preparation, delivery, and approval of drawings, specifications, and associated documentation must be clear, complete, and current before they receive final approval. The engineering department controls all parts lists, drawings, bills of material, operation sheets, and specifications used in the manufacture of a product. A request for an engineering change to these documents is made when the request for a change impacts contractually specified form, fit, or function. Changes that *do not* impact form, fit, or function include items such as change in documentation and the addition of clarifying notes to drawings.

3.13 CHANGE NOTICES

Contract change notices may be requested from any member of a CMT. (See Figure 3.3 and Figure 3.4.) At this point in time, it is merely a request for a change. When contractually specified, the customer's quality assurance or production representative will review proposed changes that are submitted by a supplier to its company for concurrence and engineering approval. New or revised drawings and engineering changes are then issued to material control, production, and quality control departments. Technical documents that are modified or become obsolete are removed from active manufacturing areas and other pertinent sites in a company by a representative of the quality or engineering department and replaced with current documents. To expedite drawing change information when delivery time of a product is an imperative, an interim change notice may be distributed to personnel who have an immediate need for this information. When this change notice is received, it is attached to the basic drawing where it shall remain until the basic drawing it is completely revised.

1. Job no.	2. Item no.	3. Drawing no.	4. Specification no.

5. Item description _____

6. Requested by _____ Date _____

7. Approved by _____ Date _____

8. Nature of recommended change

_____

9. Reason

_____

10. Parts/material on hand_____ 11. Balance on orders _____

12. Disposition: rework _____ dispose _____ accept as is _____

13. Engineering change number_____ Dated _____

14. Distribution: purchasing _____ production _____ engineering_____

     quality assurance _____ contract administrator _____

15. Drawn by _____ Date _____

16. Checked by _____ Date _____

17. Approved by _____ Date _____

Figure 3.3   Engineering change notice.

Block number	Action
1. Job number	Enter the number assigned to the production lot.
2. Item number	Enter the number referenced in the respective contract, purchase order, or job order.
3. Drawing number	Enter the number shown on the respective drawing.
4. Specification number	Enter the number referenced in the respective specification.
5. Item description	Enter a brief description of the item in question.
6. Requested by/date	Enter the name of the department manager and where the change request originated.

Figure 3.4   Instructions for preparing an engineering change notice. *(Continued)*

(Continued)

Block number	Action
7. Approved by/date	Enter the name of the person delegated the responsibility for evaluating the requested change.
8. Nature of recommended change	Enter a brief description of the nature of the change request and its impact on form, fit, or function.
9. Reason	Enter a brief description why the request was made, keeping in mind its impact on compliance with contract quality requirements.
10. Parts/material	Enter the quantity of parts and/or raw material in stock, on hand, and those in process that are affected by the requested change.
11. Balance on orders	Enter as appropriate.
12. Disposition	Check appropriate block regarding recommended decision made after engineering review is made.
13. Engineering change	a. Enter a change number starting with engineering change number and date. For example, enter *1* for the first request followed by number *2*. b. In order to have a complete history of contract performance, these numbers should be prefixed by the respective contract or purchase order number. For example, XXXXXXXX-1, XXXXXXXX-2.
14. Distribution	Identify the departments that are involved in the engineering change and check the appropriate block.
15. Drawn by/date	Enter as appropriate.
16. Checked by/date	Enter as appropriate.
17. Approved by/date	Enter as appropriate.

Figure 3.4 Instructions for preparing an engineering change notice.

3.14 QUALITY SYSTEM STANDARD

The selection of an appropriate quality system standard is predicated on the design issues. If a supplier is required to design and supply a conforming product, then he or she will select a quality system model for quality assurance that addresses the elements that relate to design, development, production, installation, and servicing. However, when a supplier is required to furnish a product to an established design, then the supplier will select a quality system model for quality assurance that relate to quality elements associated with production, installation, and servicing. The establishment and maintenance of an acceptable quality standard is a precursor to establishing a robust contractual relationship between customer and supplier.

3.15 PLANNING AND IMPLEMENTING QUALITY AUDITS

Quality audits are established and maintained to verify whether quality activities and related results comply with planned arrangements, and to determine the effectiveness of a quality system. They are performed to verify that documented information is clearly defined and free of errors and omissions.

3.16 CONTROL OF NONCONFORMING PRODUCTS AND SERVICES

When a product does not conform to applicable technical documents, it is identified as nonconforming, held for review action, processed through a preliminary review monitor and, where appropriate, through a material review board.

3.17 CORRECTIVE AND PREVENTIVE ACTION

Effective and timely corrective action of nonconforming products and services is essential to implementing a quality system. Segregation of defective items is not enough. Most important, the cause of a defect must be found and corrected. Preventive action is most effective when an organization assigns surveillance responsibilities to a customer-complaint monitor.

3.18 ABSTRACT OF CONTRACT/PURCHASE ORDER REQUIREMENTS

An abstract or highlighted copy of a contract or purchase order provides assurance to all management personnel that specified requirements were reviewed and identified.

3.19 RECORDS

Records include all of the documents required to administer contact requirements. They include, but are not limited to, quality policy and procedures, purchase, production and quality plans, pre-award survey reports, post-award audit reports, process control charts, inspection and testing results, quality audit reports, metrology records, and conferences.

3.20 MEASURING THE MANUFACTURING PROCESSES

This action is taken to assure that production processes will be carried out under controlled conditions and to verify that a product is being produced in accordance with specified requirements.

3.21 CONTROL OF DOCUMENTATION AND DATA

Action includes the control of the latest drawings, specifications, manuals, policies, processes, procedures and work instructions, as well as inspection and testing reports, product observation records, and certificates of conformance.

3.22 PACKAGING AND SHIPPING

Procedures are established to assure that:

- Items will not released for shipment until preservation, packaging, marking, and shipping documents have been inspected and accepted

- Items will not be shipped until all required certifications have been included with the items shipped

- All items will be packed in a manner that prevents unacceptable handling, damage, deterioration, or substitution

- Packaging and shipping delegated to a subcontractor will be placed with capable packaging organization

- Inventory will be inspected periodically for deterioration

3.23 SUBCONTRACTED PRODUCTS AND SERVICES

Planning is also established and maintained for quality system elements that are delegated to a qualified subcontractor. It may include one or more elements of a quality system.

3.24 CUSTOMER-SUPPLIED PRODUCT

A customer-supplied product can be an instrument or a product.

3.25 TENDER

Procedures are established to assure that suppliers tender only products and services that conform to contractual requirements and that delivered products and services are supported with objective quality evidence.

3.26 GENERAL PURPOSE INSTRUMENTS AND MEASURING STANDARDS

Measuring instruments calibrated by the supplier as well as those that are calibrated by an independent calibration laboratory are identified in a quality plan.

3.27 INSPECTION AND TEST STATUS

Production lots are accompanied with a shop routing record which shows items that are waiting for inspection, items inspected and accepted, and items that were found to be nonconforming. The quality status of the product is also identified on an accepted material tag, on an observation record, on a nonconforming material tag, in a preliminary review report, and in a material review board report.

3.28 IDENTIFYING TRAINING NEEDS

A quality plan should address the competence, education, skills, and experience that are required to fulfill a contractual obligation. It should also identify training that will be required from new employees or training that will be provided from an independent consultant.

3.29 SERVICING

Established procedures for performing, verifying, and identifying the quality of services provided to the customer are identified. In addition, where there is a need to hire new employees to support servicing requirements or for the acquisition of capable consultants, that too shall be included in the plan. (See Chapter 5, Figure 5.21.)

3.30 APPLICATION OF STATISTICAL TECHNIQUES

Where practicable, plans are implemented to measure the processes that create a product with the use of control charts.

3.31 ORIENTATION CONFERENCES

Orientation conferences (OC) are convened with representatives of the first and second parties and, where appropriate, a third party (client, supplier and subcontractor respectively). They are directed toward assuring that nothing in a purchase, manufacturing, or quality plan is left open to assumption or misinterpretation. An OC aids key personnel in achieving a clear and mutual understanding of a statement of work and related contract quality requirements. This is an imperative for performing contract quality requirements correctly the first time. Orientation conferences are normally required where:

- Information in a contract is not clear enough to make a determination regarding responsibility and application

- Items exceed latest state-of the art parameters

- Tailored requirements of an item are complex in nature and will require input from the buyer's technical specialists

3.31.1 Conference attendees should have a thorough understanding of contract quality requirements. In addition, participants should review all available records and quality history that might contribute to an established agenda. Commitments made at the time of the conference should be documented and reviewed by all participants to assure that contract requirements will be met. A commitment made by a supplier regarding open items that will require future action should also be documented. Action items resulting from a quality conference should specify which individuals are responsible for taking necessary action. Under this condition, both parties will then determine if the action required will necessitate a modification in their respective purchase and quality plans. The conferences are usually chaired by the contract administrator.

3.31.2 Subjects covered during an OC include, but are not limited to, the following:

- A contract that specifies a concurrent first-article inspection and testing between a customer's and supplier's key quality assurance representatives, but does not indicate the degree of participation by the customer's designated quality assurance representative.

- An occasion in which a specification is required for certification or qualification of personal and/or equipment, but for which the contract is silent with respect to who performs the certification or qualification actions (purchaser, supplier, or a third party).

- A contract that specifies the application of an appropriate quality system standard tailored to meet special requirements, but which is vague and/or ambiguous.

- A contract that is silent as to who has responsibility for maintaining the accuracy of a customer-owned measuring instrument furnished to the supplier. Under this condition, corrective action is imperative in order to maintain the accuracy of the instrument.

- A supplier may be requested to furnish the customer with an advanced copy of a quality plan that will require certain input by the customer. However, the contract is not clear regarding the customer's role in establishing the quality plan, the point of contact, names of participants, or knowledge of when and where required actions will take place.

- The purchaser may request a cross reference matrix from the supplier showing the relation of specified quality system requirements to the supplier's established quality plan associated with tailored contract requirements. However, specified requirements are vague and will require input from the customer's designated technical representative.

3.31.3 There are two types of orientation conferences. (See Figure 3.5.)

a. A Type 1 conference involves small organizations and participants usually involve just a representative from the buyer's organization and one from the seller's organization.

b. Type 2 conferences involves various specialists who support the efforts of a contract administrator.

3.31.4 There are many subjects that are discussed during an orientation conference. Here are a few of them:

- Identification of participants
- Contract/purchase order number
- Complexity of the product and/or service
- Urgency of the delivery schedule
- Acquisition history of similar or identical products and services previously produced by a supplier
- Time and place of the conference
- Identification of the chairperson
- Conference agenda
- Individual responsible for the preparation of conference minutes
- Timely distribution of conference minutes
- Person(s) responsible for the correction of open items
- Discussions and agreements
- Supplementary contract quality requirements
- Conclusions and recommendations

3.31.5 The individual who determines that an OC is required will identify the time and place of the conference, for prepare an agenda (see Figures 3.6 and 3.7), and notify customer and supplier participants. A designated chairperson prepares and signs the report of a conference. It covers all items discussed. This includes items requiring resolution, the names of participants assigned responsibility for recommended action items, and the due date for required actions. Copies of the report are furnished to department supervisors who have a need for this information. In deference to costs and delivery schedules, an OC conference should take place prior to a scheduled production date. There are two other reasons that usually trigger the need for conducting an orientation conference:

- For new suppliers who are involved with the production of a new product design that is complex in nature.
- Where there is a requirement for first-article inspection and testing by both customer and supplier quality assurance representatives.

3.31.6 An invitation may be extended to subcontractors to attend a quality assurance conference where appropriate. A conference may be conducted at a designated area that is convenient to participating members of the conference or it may be conducted by the electronic media. On-line conferencing or tele-conferencing are not only convenient, but also cost effective. The time, date, and detail arrangements of an OC are determined by the originator of the conference.

3.31.7 A Type 1 conference is directed toward ensuring that nothing in the area of quality is open to assumption or misunderstanding, and it is conducted by customer and supplier quality representatives.

3.31.8 A Type 2 quality assurance conference may be initiated by the purchaser of a product or service or by the supplier where there is justifiable evidence of ambiguity or the need to clarify differences of opinions associated with technical or other contractual requirements.

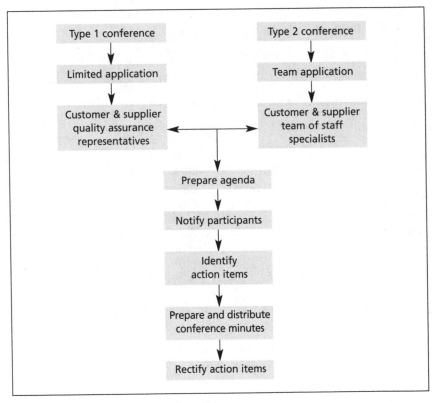

Figure 3.5 Types of orientation conferences.

1. Contract no._____ 2. Purchase order no._____

3. Conference date_____ 4. Time_____

5. Location of conference _____

_____

_____

6. Name of attendees:

   Customer_____

   Supplier_____

7. Type of conference: ☐ team application   ☐ limited application

8. Topics: First-article inspection and testing
   a. Drawings
   b. Specifications
   c. Concurrent inspection and testing
   d. Supplementary contract quality requirements
   e. Tailoring of technical requirements
   f. Documentation
   g. Delivery schedule
   h. Adopted Quality Management System
   i. Service contract

9. Report _____

_____

_____

_____

_____

_____

_____

_____

10. Chairperson _____ Title_____ Date_____

Figure 3.6 Quality assurance conference agenda.

Block number	Action
1. Contract number	Enter the contract number that is related to the conference agenda.
2. Purchase order number	Enter the purchase order number that is related to the conference agenda.
3. Conference date	Enter the date agreed upon between customer and supplier representatives.
4. Time	Enter as appropriate.
5. Location of conference	Show the address and attach directions to the conference site.
6. Name of attendees	Enter the name of key customer and supplier representatives who plan to attend the conference.
7. Type of conference	Enter a mark in the applicable block.
8. Topics	Enter agenda topics recommended by key customer and supplier representatives.
9. Report*	Enter: a. Relevant discussions, agreements, and action items. b. Open action items. c. Person(s) responsible for addressing open items. d. Recommended completion date of open items. e. Progress reports where appropriate. f. Other as appropriate.
10. Chairperson/title/date	Enter as appropriate.

Use continuation sheet where appropriate. (See Figure 3.8.)

Figure 3.7   Instructions for preparing a quality assurance conference agenda.

Continuation Sheet

Title of document _____ Document no. _____

Block no.   Comments

_____   _____

_____   _____

Prepared by _____ Title _____ Date _____

Figure 3.8   Continuation sheet – general purpose.

3.32 SOFTWARE QUALITY ASSURANCE PLAN

"The plan shall reference or document the [supplier's] procedures for preparation and execution of reviews and audits, for establishing the traceability of initial contract requirements, and for ensuring that the reviews and audits are conducted in accordance with prescribed procedures. The schedule for review and audits shall be referenced or stated in the plan."[2]

When the application of a software quality assurance program is referenced in a specification, contract and/or purchase order, it shall become a component of the overall quality plan. The purpose of the plan is to ensure that software developed under a contract complies with specified requirements. The program requires the performance of periodic audits. (See audit checklist, Chapter 5, Figure 5.23.)

3.33 RECOMMENDED ACTION

3.33.1 Documented audits should clearly identify the individual responsible for recommended actions. The director of quality assurance or his/her designated representative who participated in an orientation conference maintains surveillance of all quality assurance actions recommended and agreed upon at the conference. Customer and supplier personnel who are invited to attend the conference should be familiar with all contractual requirements before they participate in the conference. Agenda topics should be prepared in sufficient time prior to convening the conference to assure that topics to be covered are clearly understood by all, including where appropriate, subcontractor quality assurance representatives.

3.33.2 When an in-depth analysis and a clear understanding prevails between customer and supplier at a quality assurance orientation conference, it creates an enriched environment that will enhance the satisfactory implementation and maintenance of specified requirements. When agenda topics are carefully addressed and where proactive action is taken and properly implemented, benefits accrue by all parties.

3.33.3 It is important to note that the purpose of a quality assurance orientation conference is not to change contractual requirements, but to clarify specified contract quality requirements and to agree with the contents of a documented quality plan.

3.33.4 The chairperson prepares and signs the conference report, that covers all items discussed including areas requiring resolution, controversial matters, names of the participants assigned the responsibility for further action, and the due dates of action items. Failure to recognize the impact of satisfactory contract management can be a source of reactive effort that will be costly in time, morale, and reputation. Conversely, a contract and associated quality plan that is free of deficiencies and properly implemented will allow management personnel to focus on the value-added side of contract quality management.

3.34 SUMMARY

Quality plans are prepared subsequent to the receipt of a contract award and associated technical data package. Their purpose is to notify both internal and external customers and suppliers that specified quality requirements have been addressed. It is during the administrative application of contract quality requirements that planners are notified of contract change notices or the detection of nonconformity in the quality system uncovered but an auditor. When a contract is received for a new product design that is similar to one that is in current production, the planner will investigate the adequacy of a previously prepare plan to determine if the plan requires a modification or if a completely new plan will be required.

NOTES

1. ANSI/ISO/ASQ Q9001-2000: Model for quality management systems: Requirements, Clause 7.1.
2. Mil-S-52779:1974 Software Quality Assurance Program Requirements.

4

Verification of Contract Compliance

4.1 OVERVIEW

4.1.1 Verification of contract compliance associated with the acquisition of a product or service is greatly dependent on an organization's commitment to quality excellence. This is achieved with the preparation of a robust policy that describes an organization's written commitment to meet the intent of an adopted quality system standard and the preparation of complementary procedures and processes that are workable and objective. Checklists are used to document contract requirements and verify the status of each line item.

4.1.2 The concept of verifying contract compliance is predicated on the premise that:

    a. Contracts define the rights and obligations of first and second parties

    b. The supplier understands its responsibility to accumulate objective quality data before offering a product or service to the customer for acceptance

    c. The customer recognizes its responsibility to assure that contractual requirements have been met prior to accepting a product

    d. Personnel are given organizational freedom to identify and rectify problems that relate to noncompliance with specified requirements

4.2 WRITTEN POLICY

Written policy represents one of the most important areas associated with the application of contract requirements because it addresses a supplier's commitment for tendering to its clients only supplies and services that conform to a purchase agreement. Documented quality policy also provides internal and external customers not only with an organization's commitment and approach to quality but its intent to implement pertinent elements of the adopted quality system standard. Policy statements also cover contract quality requirements delegated to subcontractors. Delegation includes, but is not limited to, the acquisition of services, raw material, parts, sub-assemblies, assemblies, and pertinent elements of an adopted quality management system.

4.3 PURCHASE AND QUALITY PLANS

Purchase and quality plans are equally important as policy planning. A comprehensive purchase plan provides an organized method of buying supplies and services. The purchase plan is established to provide confidence that solicited items will be obtained from reliable sources exactly as defined in a contractual agreement. A quality plan is developed to identify how contract quality requirements will be met. This is achieved when sound management of quality assurance effort is applied and the interests of both customer and supplier prevail.

4.4 VERIFICATION PROCESS

4.4.1 The verification process is a technique used to verify that items produced and offered for acceptance do (or do not) comply with specified contract requirements before each contract line item is tendered for acceptance.

4.4.2 Verification of contract compliance starts immediately after the completion of policies, procedures, processes, and a quality plan. The plan references pertinent quality system elements and related factors associated with each new product design. A design may be one that is in the process of development or one that is already designed. Verification of the quality of supplies and services requires the preparation of procedures and processes that provide guidance to management personnel regarding the application of a statement of work.

4.4.3 Procedures and processes provide guidance regarding the detection of potential problem areas that may develop during the administrative application of a contract. Contract administrators, with support from a team of specialists, develop and apply procedures for compliance with contractual obligations. Up-front actions to be considered during the contract verification process include:

a. **Verification of Capability:** Verification of a supplier's capability is required when information is not on hand or it is not readily available to made a determination regarding responsibility. This action may require separate supplier capability determinations where a proposed supplier does not have ANSI/ISO/ASQ Q9001-2000-certified status or some other recognized standard.

b. **Orientation Conference:** A post-award orientation conference is conducted, where necessary, to assure that contract requirements are clearly understood by the producer of products and services, and to assure that contract requirements are free from ambiguity and misunderstanding.

c. **Mutual Agreement:** There must be assurance that a mutual agreement exists between buyer and seller regarding the interpretation of each contract requirement and that differences of opinions are identified before commencement of production processes, where ambiguity, difference of opinion, or misunderstanding may be present.

4.5 INSPECTION AND TESTING

4.5.1 Manpower efforts regarding the administrative application of contract quality requirements and the verification of contract compliance includes the assessment of data generated from inspection and testing. The data is used in concert with the application of procedures and processes that support the assessment of pertinent quality elements and related factors that lead to verification of contract compliance. Inspection and test data is used to verify that the adequacy of procedures and processes that support decisions to accept or reject contract line items. There are a number of reasons to conduct product inspection and testing, including:

a. Measure process capability

b. Measure process variability

c. One of the focal points of performing quality audits

d. A focal point when investigating customer complaints

e. Verify the quality status of contract line items

f. Examine product, fit and function, stated or implied

4.5.2 The quality of data accumulated through inspection and testing is dependent on the use and application of inspection, measuring, and test equipment of known accuracy. Configuration of the manufactured product will dictate whether characteristics selected for measurement and subsequently included in a quality plan will be inspected through attributes or variables sampling. Each should be carefully considered and addressed accordingly when establishing a quality plan.

4.6 VERIFICATION OF CONTRACT COMPLIANCE VIA INSPECTIONS BY ATTRIBUTES

4.6.1 *Inspection by attributes is inspection whereby either the unit of product is classified as conforming or nonconforming, or the number of nonconformities in the unit of product is counted, with respect to a given requirement or a set of requirements.*[1]

Attribute type characteristics are inspected with fixed *go, no-go* inspection gages, and they do not involve numerical values. Consequently, compliance with specified technical documents rests solely on the accuracy of fixed measuring devices. They just tell the user of an instrument if the measured characteristic is "good" or "bad." Results of attribute measurements must be documented for audit evaluation purposes and process improvement techniques. The method of documentation, hard copy or electronic, is at the discretion of a producer of a product or service.

4.6.2 The following items are addressed when recording the results of attribute sampling inspection:

 a. Item description.

 b. Specification number.

 c. Drawing number.

 d. Inspection station number. (See legend, Item u. below.)

 e. Contract or purchase order number.

 f. Sampling plan.

 g. Date.

 h. Contract change notice number.

 i. Lot size.

 j. Sample sizes.

k. Accept number specified in a selected sampling plan.

l. Rejection number specified in a selected sampling plan.

m. Master requirement list number. The applicable master requirements list code number referenced in an associated quality plan.

n. Characteristic and measuring device. The respective product characteristic and measuring device code numbers referenced in an associated quality plan.

o. Number of observed defectives characteristics.

p. Number of observations.

q. Total number of defective observations.

r. Calculated percent defective.

s. Disposition of production lot.

t. Initials or inspection stamp of the person who performed the inspection or test.

u. Legend:
No. 1 = Receiving inspection
No. 2 = First-article inspection
No. 3 = In-process inspection
No. 4 = Final inspection
No. 5 = Sampling inspection

4.6.3 A record of attributes can also used for recording other information such as:

a. The adequacy or inadequacy of certificates of conformance such as physical and chemical analysis of raw material

b. The identification of related inspection, measuring, and test equipment

c. The inspection of preservation, packaging, and marking of shipping containers

d. Traceability of inspection, measuring and test equipment and product characteristics to a specific job order, contract, or purchase order

4.6.4 The form should be designed to record inspection results of multiple lots over an extended period of time, thus providing managers and quality auditors with a complete history of a product (good or bad) on a single document.

4.7 VERIFICATION OF CONTRACT COMPLIANCE VIA INSPECTION OF VARIABLE TYPE PRODUCT CHARACTERISTICS

4.7.1 *Measurement of quality by the method of variables consists of measuring and recording the numerical magnitude of a quality characteristic for each of the units in the group under consideration. This involves reference to a continuous scale of some kind.*[2]

4.7.2 The variable inspection method determines how good or bad a product characteristic is by making and analyzing actual measurements. With variable type inspections fewer observations are necessary for determining the quality of a product. Figure 4.1 is one example of how variable data is documented. Figure 4.2 gives instructions for completing that form.

---

1. Contract/purchase order no. _____   2. Master requirements list no._____

3. Product inspection station:

    First article_____ Receiving _____ In-Process_____ Final _____

4. Part nomenclature   5. Drawing no.   6. Specification no.

_____ _____ _____

7. Drawing dim.	8. Spec. para.	9. Specified req't.	10. Observed value	11. Disposition	12. Inspector	13. Date
_____	_____	_____	_____	_____	_____	_____
_____	_____	_____	_____	_____	_____	_____
_____	_____	_____	_____	_____	_____	_____
_____	_____	_____	_____	_____	_____	_____

14. Sampling plan_____   15. Lot size _____   16. Sample size_____

17. Number of nonconforming product _____ 18. Percent defective_____

19. Prepared by_____ 20. Title_____ 21. Date_____

22. Audited by _____ 23. Title_____ 24. Date_____

Figure 4.1   Inspection record of variable type product characteristics.

Block number	Action
1. Contract/purchase no.	Enter as appropriate.
2. Master requirements list	Enter master list code number shown in a quality plan.
3. Product inspection station	Place a check mark in the block where the inspection or testing was performed.
4. Part nomenclature	Enter as appropriate.
5. Drawing no.	Enter the applicable part number referenced in the respective contract or purchase order.
6. Specification no.	Enter the applicable specification number referenced in the respective contract or purchase order.
7. Drawing dimension	Enter the numerical value referenced in the applicable drawing.
8. Specification paragraph	Enter the applicable paragraph number referenced in the applicable specification.
9. Specified requirement	a. Enter the applicable product characteristic.
	b. Where a specification is denoted and space will accommodate the product characteristic, enter it as appropriate. If not, enter a traceable code number.
10. Observed value	Enter the observed inspection or test parameter.
11. Disposition	Enter *A* for acceptable or *R* for not acceptable.
12. Inspector	Enter the inspector's initials or authorized inspection stamp after inspection or test is completed.
13. Date	Enter inspection date.
14. Sampling plan	Enter the sampling plan specified in a contract or as specified in a quality plan.
15. Lot size	Enter the number of units of product in a production lot.
16. Sample size	Enter the sample size referenced in an established quality plan.
17. Number	Enter the number of nonconforming product characteristics found in the production lot.
18. Percent defective	Enter calculated value (number of nonconforming units divided by the total number of units, multiplied by one hundred).
19. Prepared by	Enter the name of the director of quality assurance, or his or her designated quality assurance representative.
20. Title	Enter as appropriate.
21. Date	Enter the date when the audit was prepared.
22. Audited by	Enter the name of the person who audited the information entered on this form.
23. Title	Enter as appropriate.
24. Date	Enter the date when the audit was conducted.

Figure 4.2 Instructions for completing record of variable type product characteristics.

4.8 RECEIVING INSPECTION

4.8.1 Receiving inspectors work from a copy or an abstract of a contract or purchase order and applicable specifications, drawings, and job order, along with documented inspection instructions. When a shipment is found to be in compliance with a purchase agreement, the following information will be recorded. This information is usually entered on a form set aside specifically for these observations:

    a. Part name referenced in the job order folder

    b. Job order number referenced in an applicable quality plan

    c. Contract/purchase order number shown in a job order folder

    d. Inspection/rejection status

    e. Remarks (for example: *move to next operation or storage*)

    f. Inspector's signature and date

4.8.2 After acceptance, an item will either be routed to storage or routed to a production site. Items sent to production are accompanied with a shop routing record. This document remains with a product throughout subsequent manufacturing, inspection, and shipping processes.

Conversely, unacceptable production lots are identified on a red tag. The product is then routed to a nonconforming material holding area and processed through a preliminary review monitor. Where appropriate, the product will also be routed to a material review board for assessment, conclusions, and recommendations. Information referenced on a nonconforming tag includes the following:

    a. Part name referenced in the job order folder

    b. Job number shown in an applicable quality plan

    c. Reason for rejection (a brief description of the observed nonconformance)

    d. Inspector's name

    e. Date that the nonconformity was detected

4.8.3 Objective Quality Evidence

Where a certificate of conformance is requested from a supplier of raw material, the receiving inspector will accept raw material only when it is determined that certifications associated with test data and physical and chemical analysis reports are found to be acceptable. When this information is missing or is not furnished as specified in a purchase order or contract, this condition will be brought to the attention of the purchasing manager by the receiving inspector via the director of quality assurance for necessary corrective action. Without a certificate, which must be furnished with each shipment, the receiving inspector will not process the material through receiving inspection. The material will be held in abeyance until the certificate is received from the supplier. Corrective action regarding this condition is the responsibility of the director of purchasing. Validation of satisfactory corrective action taken is the responsibility of the director of quality assurance or his designated representative.

4.8.4 Customer-Supplied Product (CSP)

A customer-supplied product is one that is considered to be a product, special tooling, and/or special inspection, measuring and test equipment—acquired by a customer and contractually made available to a supplier—which may or may not be consumed in performing a contract. Product and/or measuring instruments that are furnished to a supplier by a customer are examined for count and condition by the receiving inspector. Items that are found to be acceptable are identified as CSP and routed to a storage area or directly to the production line. Customer-supplied products are identified with the following information:

a. Purchase order number

b. Item serial number (where applicable)

c. Property control number

d. Job number

e. Nomenclature

4.8.5 Item identification must be permanent, legible, tamper proof, and conspicuously marked on the product, label, or on the product's storage or shipping container. When identification is impractical due to size or configuration, the control record will reflect pertinent information. When a shipment of CSP is received in an unacceptable condition, it must be identified with a red nonconforming product tag. When a deficiency is attributed to shipping containers, the packages and packaging material used to ship a customer's product should not be discarded until after the reported deficiency is properly investigated and corrected.

4.8.6 Shipping discrepancies include:

    a. Shortages or overages within a container

    b. Incorrect part or stock number

    c. Misdirected shipment

    d. Missing or incorrect paper work

    e. Any other unacceptable condition detected at the time of the receipt of items which materially affects the serviceability or suitability for the purposes intended

4.8.7 The use of customer-supplied products is normally limited to the following conditions:

    a. The use of CSP for any purpose other than that specified in a purchase agreement is forbidden unless expressly authorized in writing from the customer. Any loss, damage, or destruction of CSP must be immediately reported to the customer.

    b. Identification of the retention time that "on-loan" CSP will be held by the supplier.

    c. That "on-loan" items will be returned to the customer immediately after the completion of a period stipulated in a purchase order.

4.8.8 Accountability

Key managers associated with the administrative application of customer-supplied products include:

    a. A contract administrator who is the point of contact with the customer regarding all matters pertaining to customer-supplied products.

    b. A quality control manager or his/her designated representative, who is responsible for periodic inspection of customer-supplied products, where appropriate.

    c. A product custodian delegated with the responsibility for maintaining the accountability of all CSP.

4.9 IN-PROCESS INSPECTION

4.9.1 In-process inspections are performed at product verification stations located throughout a manufacturing plant during fabrication and processing. Initially, first-piece inspection is performed to provide an indication as to whether or not the manufactured product meets contract quality requirements. It is then checked periodically to assure that the product flows smoothly through the production cycle.

4.9.2 In-process inspectors may be called upon to perform inspections on partially fabricated items or on a completed product. In carrying out their assigned responsibilities, inspectors are guided by established policies and procedures. These inspections are generally performed to upgrade the quality of workmanship and associated procedures and processes, and they are not always intended to serve as a medium for acceptance or rejection of a product.

4.9.3 Documentation of in-process inspections provides an indication to internal as well as external customers that product quality is being continually maintained. In-process inspection is a fertile area that provides managers with an opportunity to improve the processes that create better products and services through the application of process capability studies and the use of associated process control charts.

4.9.4 Documentation also includes a record of conforming, as well as nonconforming products and services. Another important in-process document is the Shop Routing Record. This record provides an ongoing account of production and inspection functions throughout all areas of a manufacturing process. This record contains the following information:

   a. Part name

   b. Drawing number specified in an applicable contract or purchase order

   c. Description of revision to a statement of work

   d. Revision number

   e. Job order number assigned to the production lot

   f. Specification number referenced in an associated contract or purchase order

   g. Contract number referenced in a job order folder

   h. Purchase order number referenced in a job order folder

  i. The total quantity of the production lot size that is routed throughout the production area

  j. Name and title of the person who identified the sequence of operations

  k. Name of the person who approved the sequence of operations

  l. Date that the routing record was approved

  m. Sequence of operations starting with Number 1

  n. List of operations in chronological order

  o. Starting and completion time of each operation

  p. Date that each line item was completed

  q. Identification of the employee who performed each operation

  r. Number of items accepted.

  s. Number of items rejected

  t. Comments that lead to process improvement techniques

  u. Name and title of auditor

  v. Date that the audit was conducted

4.9.5 A requirement for the application of a special process is established and maintained as a result of a requirement referenced in drawings, specifications, or other technical documents developed by a supplier or as specified in a contractual agreement between customer and supplier. (Fluorescent inspection, X-Ray, and magnetic particle inspection are three examples of a special process.) The control of processes that are accomplished by a supplier are based on a specification covering the process. They frequently require a supplier to prepare written procedures that cover the process.

4.9.6 When it is determined that a special process is a specified requirement, the applicable specification is reviewed to determine if there is a requirements for the certification of equipment, personnel, or validation. Where applicable, the supplier may elect to have its own personnel and equipment certified, or the requirement may be delegated to an independent testing laboratory. When a prime contractor (supplier) delegates the application of special process to a subcontractor with unknown capabilities, he or she must verify that the proposed supplier is capable of meeting requirements prior to issuing a purchase order.

4.10 FINAL INSPECTION

4.10.1 Final inspection and testing are performed in accordance with contractual requirements and an associated quality plan. Active as well as completed inspection and test records are filed in a job order folder and should be readily available to internal, and where applicable, to a customer's auditor. Observations of attribute and variable characteristics are documented on a form specifically designed for this action. After the acceptance of a product, a conforming tag is prepared and attached to a production lot. The product is then routed to the shipping department or to a controlled storage area. Products found to be nonconforming are identified and delivered to a designated holding area. Action is then taken to rectify observed nonconforming product characteristics.

4.10.2 When a contract specifies concurrent inspection and acceptance at source, the customer's quality assurance representative must be notified by the supplier of those production lots that are ready for inspection via a request for concurrent/source inspection. This notification is submitted to the customer within a time period agreed upon between customer and supplier quality assurance representatives. The information in this notice includes the following:

a. Date that the request was submitted to the customer's authorized quality assurance representative

b. Name of the supplier's quality assurance director or his/her designated representative

c. Name of customer's point of contact

d. Part nomenclature

e. Drawing number referenced in a job order folder

f. Specification number referenced in the job order folder

g. Name of the verification stations where the inspection and/or test will take place

h. Date and time that the inspection or testing will take place

i. Comments (for example: contract change notices, tailored drawings, and specifications that will require special attention, or any other information that will be of interest of both customer and supplier)

j. Name and title of the person who prepared the request for concurrent source inspection as well as the date

4.11 MEASURING INSTRUMENTS

4.11.1 Verification of contract compliance is greatly dependent on the establishment and application of a robust metrology program that is used to measure procedures and processes, and to provide objective quality evidence regarding compliance with stated requirements. The accuracy of measuring instruments must be controlled in all areas of a manufactured product that requires the use measuring instruments of known accuracy. Focus should be centered on those instruments required to fulfill contractual obligations. Calibration system requirements—whether implemented at a prime contractor's plant, at a designated subcontractor's facility, or at an independent calibration laboratory—must establish controls over contract related inspection, measuring, and test equipment. Calibration system standards published by the national and international community outline the individual parts of a calibration system. They provide guidance to producers of products and services and independent calibration laboratories regarding the uniform evaluation of each element of an adopted calibration system standard. The primary purpose of a calibration system is to verify product conformity and to detect nonconforming products. The successful application of a robust calibration system is predicated on the timely receipt of an abstract of contract quality and technical requirements. A producer of products and service should have a comprehensive plan for all segments of an adopted calibration system standard. The plan is based on a study of contract quality requirements, the product design, required inspection, measuring and test equipment, measuring standard requirements, and associated product and instrument parameters. It should define how contract quality requirements will be met within an overall quality system. It should also include provisions for preparing and implementing a Master Requirements List.[3] The list identifies product characteristic and its code number, as well as the measuring device used to inspect a product characteristic and its code number. This list is prepared for each product or assembly and requires modifications only when there are significant changes to a product design or other contract quality requirements. When properly implemented, this document provides positive evidence that satisfactory action has been taken to assure pertinent product parameters and associated measuring devices have been considered. In addition, when properly implemented and married with a product observation record, this document provides traceability of the quality history of product characteristics and the instruments that were used to inspect the product. Traceability steps of measuring instruments are as follows:

a. Product characteristic to general purpose instruments (GPI)

b. GPI to a transfer standard (TS)

c. TS to a primary standard (PS)

d. PS to the National Institute of Standards and Technology or an International Standard

4.11.2 Calibration procedures used for calibration and the repair of measuring instruments should be outlined in an instruction manual. Most important, they should be readily available to the metrology technician during initial calibration, as well as during the re-calibration of instruments for general usage and associated measurement standards. Procedures that are used for the calibration of measuring instruments are usually outlined in the instrument manufacturer's instruction manual. However, there will be situations where these procedures will not be readily available to a supplier or an independent laboratory. A producer of a product or an independent calibration laboratory will prepare its own calibration procedures in its own facility when it is determined that they are not available from the manufacturer of an instrument.

4.12 HANDLING THE PRODUCT

Items, including measuring instruments and a product, must be handled in a manner that will protect them from physical and mechanical damage when they are moved within the supplier's facility or to a customer's specified destination. An inspection checklist (see Figure 4.3) should be used when inspecting items that ready for shipment to specified destination.

Characteristic	Code number*
Preservation and packaging method	S1
Quantities	S2
Cleaning process	S3
Preservation of the product	S4
Dunnage	S5
Product protection	S6
Shipment weight	S7
Container markings	S8
Destination markings	S9
Stock number	S10
Item description	S11
Serial numbers (where applicable)	S12
Special testing, such quick leak, heat seal, cyclic exposure, or rough handling	S13
Shipping documents	S14
Submit notice to the billing department of an executed shipment	S15
Other (as appropriate)	S16

* Note: Code numbers are used in deference to the available space on a respective inspection and test report form.

Figure 4.3 Preservation, packaging, and shipping checklist.

4.13 CONCLUSIONS

Policies and procedures are implemented to ensure that supplies and services acquired under contract conform to a contract's quality and quantity requirements, including inspection, acceptance, and other measures associated with contract requirements. A provider of products and services is responsible for tendering to its client for acceptance only those supplies and services that conform to a contractual agreement.

NOTES

1. ANSI/ASQ Z1.4 2003: Sampling Procedures and Tables for Inspection by Attributes, ASQ.
2. *Glossary and Tables for Statistical Quality Control,* Clause 137. Milwaukee, WI: ASQ Quality Press, 2004.
3. *Managing the Metrology System,* 3rd ed. Milwaukee, WI: ASQ Quality Press, 2004.

5

Audit of Contract Compliance

5.1 BACKGROUND

5.1.1 *An audit [program] shall be planned, taking into consideration the status and importance of the processes and areas to be audited, as well as the results of previous audits. The audit criteria, scope, frequency, and methods shall be defined. Selection of audits and conduct of audits shall insure objectivity and impartiality of the audit process. Auditors will not audit their own work.[1]*

5.1.2 Audits provide assurance to both internal and external customers that contract requirements are clearly defined and that established policy, procedures, and processes are (or are not) achieving defined objectives. The following are basic elements of a documented quality system that an auditor looks for at the outset of an audit:

- That policy is established for all of the elements of a pertinent quality system

- That required operation procedures and processes are documented, in place and properly applied

- Top management's support to his or her management and operations personnel

5.1.3 Organizations that take a proactive approach regarding the application of an audit program want assurance that:

- Audits are objective and free from bias

- Audit will lead to continuous improvement of procedures and processes

- Audits are performed by qualified individuals

5.1.4 Advance knowledge of what administrative and technical requirements are imposed in a contract is essential when selecting pertinent audit factors. A robust audit is one that centers its focus on the evaluation of product parameters and systemic factors that are most representative of a contractual requirement.

5.2 AUDIT PROCESS

5.2.1 The basis for auditing policies, procedures, and processes are the contract related checklists that are prepared during a desk audit of a planned audit. It is at this point that a determination is made as to whether all policies, procedures, and processes have been considered, that they are in place, and they are properly documented. When it is determined during the desk audit that certain documentation is found to be incomplete or missing, an organization may elect to defer an audit until satisfactory corrective action is taken to rectify observed conditions. Or an organization may elect to have the auditor proceed with the audit to determine if existing procedures are implemented in a satisfactory manner and then make appropriate conclusions and recommendations.

5.2.2 Customers audit their own contract policies and procedures prior to soliciting the services of suppliers of products and services to assure that their needs and expectations will not be overlooked or compromised.

5.3 AUDIT CHECKLISTS

In deference to protecting its own needs and interests, the purchaser (first party) audits its supplier (second party) and the second party audits its subcontractors (third party). These audits should always be contract specific, objective, and meaningful. First-party audit checklists should be directly related to functions performed by the second party. To preclude duplication of effort and unnecessary loss of man hours, second-party audits performed at a third party's facility should relate only to the applicable portions of contract quality requirements that are delegated to them.

5.3.1 Audit checklists that are used repeatedly may be modified or improved subsequent to:

- Feedback recommendations made by the owner(s) of a process

- Internal and external customer complaints

- Nonconforming quality reports

5.3.2 Auditors must be unbiased and be familiar with policy, procedures and processes. (See Figure 5.1.)

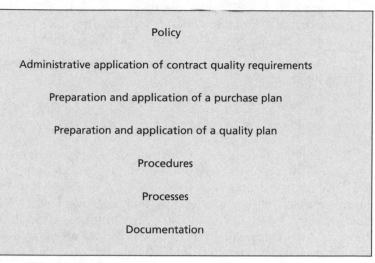

Policy

Administrative application of contract quality requirements

Preparation and application of a purchase plan

Preparation and application of a quality plan

Procedures

Processes

Documentation

Figure 5.1   Administrative functions of contract quality management.

5.4 AUDIT TOPICS TO CONSIDER

Topics to consider when performing a contract-related audit include the following:

- Distribution of an audit schedule to all participants
- Selection of one or more product and/or service characteristics scheduled for evaluation
- Identification of the procedures that will be examined
- Identification of components of purchase, production, and quality plans
- Identification of administrative and product/service verification stations scheduled for an audit
- Identification (where applicable) of deficient product characteristics and quality system elements observed during a previous audit
- Conduct a pre-audit conference with the team leader and other members of an audit team to assure that there is a clear and objective understanding of agenda topics
- Copy of the audit agenda to participating supervisors

- Assurance that the audit will be conducted in an objective and unbiased manner

- Identify the reason for conducting the audit, for example, internal or external customer complaints, quest for process improvements, or verification of quality assurance capabilities

- Availability of policies, procedures, contractual and technical documents

5.5 WHY AUDITS ARE IMPORTANT

5.5.1 Audits are designed to examine the ability of established procedures and processes to consistently produce conforming products and services requirements and to identify opportunities for their improvements. These opportunities and improvements are gained when quality assurance specialists who conduct an audit have a comprehensive knowledge of the procedures and processes associated with the design, development, production, installation, and servicing of a product. There are many benefits that are realized as the result of a quality audit. For example:

- They establish an objective relationship between first and second parties

- They provide assurance to their respective organizations that needs, interests, and expectations are being met

- They assure that defined responsibilities are implemented correctly the first time

- They assure that specified contract requirements are adequately implemented and maintained throughout all phases of a production cycle

- They determine that key plans are established and implemented properly

- They determine that a supplier is measuring processes effectively

- They determine whether or not a supplier is making unbiased decisions when accepting or rejecting a product

- They support the objectives of statistical process control techniques

- They provide confidence that satisfactory procedures and processes are in place

- Most important, they identify an organization's strengths and weaknesses

5.6 QUALITY AUDITS

Quality audits must be performed by someone who is familiar with contractual requirements: the applicable quality system standard, associated technical requirements, and related policies and procedures.

5.7 DEGREE OF AUDIT INPUT

The degree of audit input is predicated on the complexity of a product design and complexity of a process. Hence, the audit of products and service requirements that are complex in nature, unlike noncomplex items, requires an extensive amount of planning and implementation. Each should be carefully considered during the establishment of an audit plan. Where a purchase involves items that are noncomplex in nature, planners should take into account the degree in which an audit will be implemented or whether an audit is really necessary.

5.8 AUDIT METHOD

5.8.1 In deference to objective and cost-effective planning, a determination should be made as to whether the application of a *macro* or *micro* audit (see Figure 5.2) will be in the best interest of an organization. Where an organization is involved in sporadic production, the application of a micro audit is recommended. Conversely, the application of a macro audit applies where the organization is involved with continuous quality assurance activity that involves many policies, procedures, and processes.

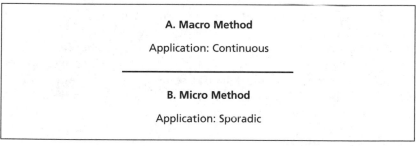

A. Macro Method

Application: Continuous

B. Micro Method

Application: Sporadic

Figure 5.2   Audit methods.

5.9 IMPLEMENTATION

5.9.1 Audit of contract quality requirements and associated policy, procedures, and processes are performed under following conditions:

- Prior to and after the ANSI/ISO/ASQ Q9000 – 2000 certification and registration process

- Prior to and after quality assurance capability is established

- Prior to and after first-article inspection and testing

- During a suppliers quest for achieving continuous improvements of systemic related policies, procedures, and processes

- During product inspection and testing

- During the application Software Quality Assurance programs (SQA), as referenced in an item specification, contract, or a purchase order

5.10 MACRO METHOD

5.10.1 An initial audit of a quality system under the macro method is normally performed in minute detail and on scheduled or unscheduled interval thereafter. Its focus is centered on quality system elements and product and service characteristics that have a direct impact on the quality of a delivered product or service. Audits may be conducted on all quality system elements or they may be accomplished at verification stations and related systemic factors that are randomly selected by an auditor. They should be scheduled in a manner that is convenient to all participants and in a manner that will not impede the timely performance of other operations within an organization.

5.10.2 The degree of implementation of a macro audit is predicated on the immediate needs and interests of both customer and supplier or as dictated by a contractual agreement. In deference to objective quality management and meeting the needs and expectations of both customer and supplier, a concerted effort should be made by an auditor to select product, service, and systemic factors that are objective and related to specific contract quality requirements.

5.10.3 The macro method of auditing applies to those organizations that are involved with the continuous production of goods and/or services that involve many *more* checklists, procedures, processes, and man-hours than a micro audit. Listed below are checklists that can be benchmarked against appropriate contract quality requirements (see Figure 5.3 through Figure 5.22). These checklists are generic and are intended to be used only as a benchmark when establishing an appropriate audit checklist. They should relate to specific contract requirements and are not to be stereotyped. Selected factors shall be objective and free of bias and redundancy.

5.11 AUDIT CHECKLISTS

5.11.1 Element: Management Responsibility

FACTORS:

- Commitment to quality excellence by top management
- Quality control is a separate quality control functional unit
- Leadership provided by top and middle management
- Establishment of purchase and quality plans
- Management support in problem solving
- Improvement of processes that create products and services
- Contract review and implementation
- Drawings, specifications, and change control
- Identification of product and service verifications stations
- Work instructions
- Inspection, measuring, and test equipment
- Monitoring of internal and external customer complaints
- Periodic review of policy and procedures
- Organizational structure
- Prevention of nonconforming products and services
- Freedom and authority to make proactive recommendations
- Identification of systemic problems
- Initiate, recommend, and provide problem solutions through designated channels
- Timely correction of unsatisfactory conditions
- Procedure for reporting audit results to upper management

Figure 5.3   Management responsibility checklist.

5.11.2 Element: Quality System

FACTORS:

- Quality system regarding the intent of a specified quality system standard is maintained
- Written procedures defining quality related functions
- Policy is consistent with a prescribed quality system standard
- Original manual and changes thereto are reviewed and accepted by a designated representative of top management
- Quality system implemented as documented
- Commitment to establish a quality plan for each new product design
- Quality system implemented in accordance with documented policy and procedures
- Distribution of quality policy

Figure 5.4   Quality system checklist.

5.11.3 Element: Contract Review

FACTORS:

- Review
- Summarize
- Document
- Adequately define
- Distribute an abstract of contract quality requirements
- Distribute contract changes and amendments
- Review and distribute contract amendments in a timely manner

---

Figure 5.5   Contract review checklist.

5.11.4 Element: Design Control

FACTORS:

Design and Development Planning
  a. Plan developed for each design and development activity
  b. Defined responsibilities implemented
  c. Documented procedures established and maintained
  d. Qualified personnel equipped with adequate resources

Organizational and Technical Interfaces
  a. Interfaces between different groups defined
  b. Required information is transmitted and regularly reviewed

Design Input
  a. Design-input requirements identified and documented
  b. Design-input reviewed for adequacy
  c. Incomplete, ambiguous, or conflicting requirements resolved
  d. Associated contract-review activities taken into consideration

Design Output
  a. Design output validated against design-input requirements
  b. Meets the design-input requirements
  c. Characteristics that are crucial to safety and functioning of the product identified
  d. Design-output documents reviewed before release

Design Review
  a. Formal reviews are planned and conducted
  b. The review includes representatives of all functions concerned with the design stage
  c. Records of design reviews and design verification established and maintained

Design Verification
  a. Performed to ensure that the design-stage output meets the design-stage input requirements
  b. Design-verification measures are recorded
  c. Design validation performed to ensure that the product conforms to users defined needs and requirements

Design Changes and Modifications
  a. Identified
  b. Documented
  c. Reviewed
  d. Approved by authorized personnel before implementation

Figure 5.6 Design control checklist.

5.11.5 Element: Document and Data Control

FACTORS:

- Documents reviewed and approved for adequacy prior to issue
- Identification of document revision status
- Availability of pertinent documents at locations where functions of the quality system is performed
- Removal of obsolete/outdated documents
- Duration that outdated and obsolete drawings is to be retained
- Person/organization responsible for document change review and approval
- Nature of changes identified in the document or attached thereto
- Assurance that the latest change notices and specifications are in use by operating and quality personnel during the production process

---

Figure 5.7 Document and data control checklist.

5.11.6 Element: Purchasing

FACTORS:

- Verification of finance, production, and quality assurance capability
- Identification of quality assurance activities delegated to a subcontractor
- Maintenance of supplier quality history
- Purchasing documents clearly describe product/service ordered
- Review of purchase orders by a representative of the quality control department

---

Figure 5.8 Purchasing checklist.

5.11.7 Element: Control of Customer-Supplied Product and Measuring Instruments

---

FACTORS:

- Examined for count and condition upon receipt
- Identified as acceptable prior to routing to stock area or a production line
- Inventory control
- Periodic inspection for damage or mishandling
- Feedback procedure regarding product that is damaged, lost, or found to be unsuitable for use
- Maintenance of records

Figure 5.9   Control of customer-supplied product checklist.

5.11.8 Element: Product Identification and Traceability

FACTORS:

- Traceable to a contract or purchase order
- Traceable to a job order number
- Traceability of product and service characteristics
- Traceability of inspection, measuring, and test equipment used to check product characteristics
- Traceability of measuring standard used to calibrate instrument used to check the product
- Traceability of physical and chemical analysis of raw material

---

Figure 5.10   Product identification and traceability checklist.

5.11.9 Element: Process Control

FACTORS:

- Procedures defining the manner of production, installation, and servicing
- Compliance with specified standards, drawings, specifications, quality plans, and documented procedures
- Monitoring and control of process parameters and product characteristics
- Approval of processes and equipment
- Maintenance of equipment to ensure continuous process capability
- Application of special processes (radiography, magnetic particle inspection, welding)
- Pre-qualification of special processes
- Qualification of processes, production, equipment, and personnel
- Monitoring the status of methods, materials, and machines
- Statistical techniques

Figure 5.11   Process control checklist.

5.11.10 Element: Inspection and Testing

FACTORS:

Procedures for Incoming Inspection and Testing
- a. Incoming product conforms to specified requirements
- b. Specified requirements are accomplished in accordance with an established quality plan and documented procedures
- c. Controls implemented at subcontractor's facility
- d. Documented evidence of inspection and testing

Procedures for In-process Inspection and Testing
- a. Product inspected and tested in accordance with quality plan and documented procedures and processes
- b. Product inspection and test is accomplished prior to routing the product to the next operation

Procedures for Final Inspection and Testing
- a. Product inspected in accordance with an established quality plan and documented procedures and processes
- b. The quality plan and documented procedures are all inclusive of contract quality requirements
- c. The quality plan and procedures are readily available and clearly understood

Identification of Accepted Production Lots
- a. Identification of rejected lots
- b. Segregation of unidentified production lots
- c. Identification and control of raw material
- d. Availability of objective quality data
- e. Availability of acceptable certificates of compliance
- f. Procedure for the validation of certificates of compliance

Figure 5.12 Inspection and testing checklist.

5.11.11 Element: Control of Inspection, Measuring, and Test Equipment

---

FACTORS:

- Responsibilities for controlling the calibration status of all active and inactive
- Accuracy of all active measuring instruments
- Environmental controls
- Intervals of calibration (hardware and software)
- Control of out-of-tolerance conditions
- The identification of significant out-of-tolerance conditions
- Calibration procedures
- Calibration sources (domestic)
- Calibration sources (international)
- Application of records
- Availability of records
- Calibration status
- Handling, storage, and transportation
- Control and application of computer software
- Control of customer – supplied measuring instruments
- Inventory control of active and inactive measuring instruments
- Identification of accuracy ratios between measuring standards and general purpose instruments
- Identification of accuracy ratios between general purpose instruments and product characteristics tolerances
- Housekeeping

Figure 5.13 Control of inspection, measuring, and test equipment checklist.

5.11.12 Element: Inspection and Test Status

FACTORS:

- System for the identification of the inspection and test status
- Method of identifying inspection and test status of units and/or lots not yet inspected/tested
- Identification of unit/lots inspected and accepted
- Identification of unit/lot inspected and rejected

Figure 5.14   Inspection and test status checklist.

5.11.13 Element: Control of Nonconforming Product

FACTORS:

- Segregated
- Clearly identified
- Preliminary review
- Material review board
- Authorization to reworked the product to meet specified requirements
- Authorization to repair the product to meet specified requirements
- Request for waiver approval

Figure 5.15   Control of nonconforming product checklist.

5.11.14 Element: Corrective and Preventive Action

FACTORS:

Corrective Action
- Customer complaint monitor
- Record of internal and external customer complaints
- Tracking justified complaints
- Identification of cause(s)
- Adequacy of corrective action decisions

Preventive Action
- Maximum use of internal and external customer complaints and audit results
- Objective quality records
- Management support of corrective action processes
- Internal and external quality audits
- Use of measuring instruments of known accuracy
- Formal and documented corrective action plan (customer complaint monitor)

Figure 5.16   Corrective and preventive action checklist.

5.11.15 Element: Handling, Storage, Packaging, Preservation, and Delivery

---

FACTORS:

- Documented inspection and testing of packaging and preservation methods
- Methods for handling containers
- Prevention of damage of the product during handling
- Designation of storage areas for the product
- Appropriate methods for authorizing release of the product from storage
- Periodic inspection of items in storage
- Control of packing, packaging, and marking to assure conformance to specified requirements
- Verification that packing, packaging, and preservation requirements are in accordance with specified requirements
- Method of preservation
- Product protection after final inspection and test
- Protection during transportation
- Handling and distribution of shipping documents

Figure 5.17   Handling, storage, packaging, preservation, and delivery checklist.

5.11.16 Element: Control of Quality Records

FACTORS:

- Legible and readily available
- Reviewed by management periodically
- Records demonstrate conformance/nonconformance to specified requirements
- Records cover all purchase and quality plan requirements
- Duration that completed records are held in storage
- Protection of records while in storage
- Availability of records applicable to internal and external customers

---

Figure 5.18   Control of quality records checklist.

5.11.17 Element: Internal Audits

> **FACTORS:**
>
> - Availability of an audit plan
> - Audit schedule
> - Audit priority
> - Performed by someone other than the owner of a process
> - Method of recording audit results
> - Distribution of audit results to responsible managers
> - Person responsible for corrective action on reported deficiencies
> - Follow-up action regarding open items
> - Audit results reviewed by top management

Figure 5.19   Internal quality audit checklist.

5.11.18 Element: Training

> **FACTORS:**
>
> - Documented training program
> - Training monitor
> - Employees record of previous training
>   a. List of skills accomplished
>   b. Number of years of experience within a specific discipline
> - Identification of employee training needs
> - Scheduled training areas:
>   a. Certified quality engineer
>   b. Certified reliability and maintainability engineer
>   c. Certified quality auditor
>   d. Certified quality manager
>   e. Certified mechanical inspector
>   f. Certified quality technician
>   g. Applied statistical process control
>   h. Training for assigned task
>   i. Certified calibration technician
>   j. Other

Figure 5.20   Training checklist.

5.11.19 Element: Servicing

FACTORS:

- Verification that servicing meets specified requirements and customer satisfaction
---

Figure 5.21   Servicing checklist.

5.11.20 Element: Sampling Techniques

FACTORS:

- Procedures for the application of statistical process controls
- Selection of valid sampling plans
- Application of sampling plans
- Supplier-designed sampling plan
- Variable type sampling plan
- Attribute type sampling plan
- Selection of samples from product lots
- Process control charts
- Lot-by-lot sampling
- Analysis of quality data
- Traceability of samples

Figure 5.22   Statistical process control technique checklist.

5.11.21 Element: Software Quality Assurance (SQA) Program Requirements[2]

FACTORS:

Plan Documentation
- Organizational responsibilities
- Authorities for execution of the SQA plan
- Events critical to the implementation of the plan

Tools, Techniques, and Methodologies
- Systems analysis techniques
- Function and performance requirements analysis
- Error analysis
- Specification tracing
- Coding conventions

Computer Program Design
- Design logic
- Fulfillment of requirement
- Completeness
- Compliance with specified standards
- Design documentation subject to independent review prior to its release for coding

Work certification
- Description
- Authorization
- Completion of work performed under contract

Documentation
- Standards
- Practices
- Delivery of documentation and change information

Computer Program Library Controls
- Documented procedures and controls for the handling of codes and related data
- Assurance that different computer program versions are accurately identified and documented
- Assurance that approved modifications are incorporated
- Assurance that software submitted for testing is the correct version

---

Figure 5.23   Software quality assurance program checklist.   *(Continued)*

(Continued)

Reviews and Audits
- Documented
- Procedures for preparation and execution of reviews and audits
- Traceability of initial contract requirements
- Reviews and audits are conducted in accordance with prescribed procedures
- Schedule for review and audits are references in the plan

Configuration Management (CM)
- Assurance that the objectives of the CM program are being attained

Testing
- Analysis of software requirements to determine testability
- Review of test requirements and criteria for adequacy, feasibility, traceability, and satisfaction of requirements
- Review of test plans, procedures, and specifications for compliance with authorized changes
- Verification that tests are conducted in accordance with approved test plans and procedures
- Certification that test results are the actual findings of the test procedure
- Review and certification of test reports
- Test related media and documentation are maintained to allow repeatability of tests
- Software and computer hardware to be use to develop and test software under contract are acceptable to the customer

Corrective Action
- Prompt detection, documentation, and correction of software problems
- Reporting of problems and deficiencies to appropriate management levels
- Analysis of data and deficiency reports to determine their extent and causes
- Analysis of trends to prevent the development of noncompliant products
- Review of corrective action measures to ensure that problems and deficiencies have been resolved and correctly reflected in applicable documents

Control of Subcontractors
- Policy and procedures provide assurance that all software acquired from subcontractors comply to the applicable requirements of an applicable contract

---

Figure 5.23   Software quality assurance program checklist.

5.11.22 Element: General Requirements for Risk Management

FACTORS:

Processing Hazardous Materials

- Documented processes for hazardous materials related to specified specifications or technical documentation are traceable to a contract requirement
- During the contract review process, the supplier determines if hazardous materials requirements apply
- Supplier carefully reviews contracts and specifications covering special processes to determine if there is a need for certification of equipment or qualifications of personnel
- Policies and procedures are in place for the acquisition of hazardous materials
- Procedures are in place for the delivery of hazardous materials
- Procedures identify personnel who are responsible for appraising employees of all hazards to which they may be exposed
- Relative symptoms and emergency treatment are identified where appropriate
- Proper conditions and procedures are in place for safe use of hazardous materials
- When certification or qualification requirements are subcontracted, the supplier of hazardous materials assures that the vendor or subcontractor performing the service meets the certification/ qualification requirements
- Requirements for hazardous materials are clearly defined in purchase orders to subcontractors and vendors
- A record of objective quality evidence associated with certifications and qualifications is maintained and made available to internal and external auditors

Management Responsibilities

- Policy for determining acceptable risk is defined
- Adequacy of provisions ensured
- Trained personnel assigned
- Result of risk management activities identified
- Coordination of risk management system with management and operations personnel
- Adequacy of risk management team involved with design, development, manufacturing, quality, sales, and marketing
- Risk analysis coordinated with a team whose members are involved with the development of special characteristics, failure mode effective analysis (FMEA), and the establishment of actions to reduce potential failure modes with high risk priority number
- Maintenance of risk management files

Figure 5.24 General requirements for risk management checklist. *(Continued)*

Quality Management System

- Documented policy is in accordance with requirements referenced in BS EN ISO 14971: Medical Devices. Application of risk management to medical devices and other specified contract quality requirements
- Written procedures define contract-related functions
- Policy statements are consistent with BS EN 14971:2001
- Procedures associated with the applicable quality system standards are documented
- Management's commitment to the establishment and application of a plan for new product designs containing hazardous items
- The quality system is implemented in accordance with established policy and related procedures

Contract Review

- Contract review procedure associated with hazardous material is documented
- Contract and purchase orders associated with hazardous material are summarized
- Contract requirements are prepared and distributed
- Contract change notices associated with hazardous material reviewed and distributed

Figure 5.24 General requirements for risk management checklist.

5.12 DESK AUDIT

A successful audit is directly related to the selection of objective checklists that are prepared during the desk review. This *table setting* process should be given careful consideration before conducting an on-site audit. The documented desk report is reviewed and approved prior to implementation by a person or persons who are familiar with contract quality requirements as well as associated policy and procedures. (See Figure 5.25 and Figure 5.26.) This record identifies product and systemic factors that will be used in the actual audit and it becomes an integral component of the final report.

5.13 ON-SITE AUDIT

An on-site audit report identifies areas of compliance in the area audited as well as areas of noncompliance along with conclusions and recommendations. Where appropriate, planned follow-up action will also be included in the report. (See Figure 5.27 and Figure 5.28.) Results of a quality audit may be recorded in a computer database, on a hard copy, or a combination of the two.

1. Location of facility audited			
2. Product identification			
3. Contract/purchase order no.			
4. Verification stations			
5. Quality system standard			
6. Product characteristics			
7. System element	8. Section	9. Paragraph	10. Comment
_____	_____	_____	_____
_____	_____	_____	_____
_____	_____	_____	_____
_____	_____	_____	_____
_____	_____	_____	_____
_____	_____	_____	_____
_____	_____	_____	_____
_____	_____	_____	_____
_____	_____	_____	_____
_____	_____	_____	_____
_____	_____	_____	_____
_____	_____	_____	_____

Figure 5.25 Desk audit report. *(Continued)*

(Continued)

11. Observations

_____

_____

_____

_____

_____

_____

_____

_____

_____

_____

_____

_____

_____

_____

_____

12. Auditors   13. Title

14. Reviewed by   15. Title   16. Date

17. Approved by   18. Title   19. Date

Figure 5.25   Desk audit report.

Block number	Action
1. Audit location	Enter as appropriate (supplier or sub-supplier).
2. Product identification	Enter as appropriate.
3. Contract/P.O. number	Enter where appropriate.
4. Verification stations	Enter verification station title or identification number.
5. Quality system standard	Enter the standard that is specified in a contract.
6. Product characteristics	Enter product characteristic(s) scheduled for evaluation.
7. System element	Enter the systemic elements selected for evaluation.
8. Section	Enter the section number shown in the supplier's policy manual.
9. Paragraph	Enter the paragraph number referenced in the applicable document.
10. Comment	Enter comment as appropriate, such as:
	a. See continuation sheet attached.
	b. See checklist attached.
11. Observations	Enter comments such as:
	a. Quality procedures are considered acceptable; proceed with the audit.
	b. Procedures are unacceptable.
	c. Discontinue the audit.
	d. Audit will be rescheduled, after satisfactory corrective action is taken.
12-13. Auditors/title	Enter as appropriate.
14-16. Reviewed by	Enter as appropriate.
17-19. Approved by	Enter as appropriate.

Figure 5.26   Instructions for preparing desk audit report.

1. Identification number

2. Date prepared

_____   _____

3. Location of facility

_____

4. Key persons contacted

_____

_____

5. Quality system standard

_____

6. Metrology system standard

_____

7. Supplementary contract quality requirement

_____

8. Product description

_____

_____

9. Product characteristic

_____

_____

_____

10. Quality system element	Related factors	Page	Sect.	Para.

Figure 5.27   On-site audit report.   *(Continued)*

(Continued)

11. Audit schedule

_____

_____

_____

12. Product/service verification stations

Number   Location

_____

_____

_____

_____

_____

_____

_____

_____

13. Substantiation of conclusions and recommendations

_____

_____

_____

_____

_____

_____

14. Audited by   Title   Date

_____   _____   _____

15. Reviewed and approved by   Title   Date

_____   _____   _____

NOTE: Where necessary, use general purpose continuation sheet

Figure 5.27 On-site audit report.

Block number	Action
1. Identification number	Enter alpha/numerical number that identifies the report.
2. Date	Enter date that the report was prepared by the auditor.
3. Location of facility	Enter the name of audited organization.
4. Key persons contacted	Enter the names of the responsible managers contacted at the respective product/systemic verification stations.
5. Quality system standard	Enter customer specified quality system standard. For example: ANSI/ISO Q9001-2000: Quality Management Systems: Requirements.
6. Metrology system standard	Enter customer specified metrology system standard. For example: ISO 10012-1: Quality Assurance Requirements for measuring equipment-Part I: Metrological Confirmation System for Measuring Equipment.
7. Supplementary contract quality requirement	Enter brief description of specified supplementary quality requirement.
8. Product description	Enter the name of the product, associated drawing, or specification number.
9. Product characteristic	Enter the selected product characteristic that will be a component part of the audit.
10. Quality system element	Enter the title of the selected quality element, related page, and paragraph number. (Where required, use Continuation Sheet: General Purpose: See Figure 3.8.)
11. Audit schedule	Enter the scheduled time and date that the audit will take place. Note: Participants shall be notified well in advance of the scheduled audit.

Figure 5.28   Instructions for completing the on-site audit report.   *(Continued)*

(Continued)

Block number	Action
12. Product/service verification stations	Enter the product or service station, title, identification number, and location. This number is usually indicated on a supplier's floor plan.
13. Substantiation of conclusions and recommendations	Enter the following as appropriate: a. Substantiate conclusions and recommendations with factual data that can be traceable to a specific contract or a quality requirement. b. Recognize objective quality performance. c. Identify procedures and documentation that are fertile for improvement. d. Identify policy, procedures, and processes that are missing, deficient, or not followed. e. Identify prevailing bias treatment of established policy, procedures, and processes. f. Attach a copy of the audit checklist to the audit report. g. Notify recipients of the report within a reasonable time. (Within seven working days is perceived to be reasonable.) h. When a determination is made that corrective action is required, the recommended date for corrective action shall be included in the final audit report. It is important to note that if the corrective action process requires an extended period of time for satisfactory action, progress reports must be considered as part of the corrective action process.
14. Audited by	Enter the name, title, and date of the person who conducted the audit. If the audit is a team effort, enter the chairperson's name.
15. Reviewed and approved by	Enter the name and title of the person who approved the report and indicate the date of signing.

NOTE: Where more than 30 days will be required to correct open items, and where periodic progress reports are requested by an audit team, it is recommended that a major milestone schedule be established and maintained to track all open items.

Figure 5.28 Instructions for completing the on-site audit report.

5.14 MICRO METHOD

A *micro audit* is a modified version of a *macro audit,* which is long term, more systemic, usually performed by a contract quality management team, and applied where there is continuous production of products and services. Conversely, micro audits are contract specific; they are used to support a decision to accept (or reject) a product or service characteristic offered for acceptance. The main purpose of a micro audit is to verify that a supplier is fulfilling his or her contractual obligation and to reduce or eliminate end-item inspection. Micro audits are performed by one individual, usually a quality assurance specialist. An example of a micro audit checklist includes, but is not limited to, the factors shown in Figure 5.29.

Micro Audit Checklist – General	
Selection of product samples	Identification of nonconforming products and services
Availability of work instructions	Documentation of nonconformity
Availability and adequacy of an applicable quality plan	Documentation of corrective action taken
Availability of inspection and test records	Adequacy of corrective action as related to causes of nonconformity
Availability and accuracy of inspection, measuring, and test equipment	Follow-up of corrective action (as appropriate)
Availability and source of measuring standards	Drawing/specification in is accordance contract requirement
Traceability of products characteristics	Removal of obsolete documents from point of issue and use
Traceability of domestic and/or international measuring standards	Procedure for the control of customer-supplied products
Source of instrument calibration procedures	Customer supplied products examined upon receipt
Availability of calibration procedures	Performance of periodic internal audit
Adequacy of calibration procedures	Method of product preservation
Record of calibrations	Packing and packaging materials
Record of certificates of compliance	Inspection of packing, packaging, and marking
Responsibility for controlling customer-supplied products	Verification of physical and chemical analysis reports
Identification of instruments that are significantly out-of-tolerance	Method of identification of raw materials
Description of what constitutes a significantly out-of-tolerance	
Record of inspection and testing	
Reliability of inspection and testing instructions	

Figure 5.29 Micro audit checklist – quality assurance.

5.15 AUDIT ASSESSMENT

Where appropriate, audits shall include the assessment of financial, production, and service capabilities as a component of a macro or micro audit. Examples of related checklists are shown in Figures 5.30, 5.31, and 5.32.

5.15.1 Element: Production

FACTORS:

Production Control System Is/Is Not Operational
- Historical effectiveness
- Sufficiently effective to meet production control requirements

Plant Facilities
- Adequacy of square footage under roof
- Adequacy of total space for manufacturing
- Available space for solicited products and services

Production Equipment
- Quantity required for solicited products and services
- Condition of quantity on hand (good, fair, or poor)
- Purchased equipment (type, number, and verified delivery dates)

Purchased Parts and Materials
- Description of items
- Source of subcontracted items
- Verification of delivery dates

Subcontracting (Manufactured Items)
- Description of delegated processes
- Source of subcontracted items
- Verification of delivery dates

Personnel
- Number on board
- Number of additional personnel required
- Acquisition source

Figure 5.30 Micro audit checklist – production.

5.15.2 Financial Factors

Financial Position
- Working capital
- Liabilities
- Net worth

Financial Arrangements
- Use of own resources
- Use of bank credits
- Other

Business Relationship
- Comments of proposed supplier's bank
- Track record
- Comments and reports (example: Dun and Bradstreet, Standard and Poors Credit rating)

Sales
- Current backlog of sales
- Anticipated additional dollar sale

Figure 5.31 Micro audit checklist – financial.

5.16 SERVICES

FACTORS:
- Documentation of service policy
- Record of additions, revisions, and deletions to policy
- Documented procedures and processes that create better service
- Customer complaint monitor
- Record of nonconformity
- Feedback program that measures customer satisfaction
- Service meets the needs and expectations of the client
- Method for improving customer satisfaction
- Client supplier items identified and entered into the inventory system
- Method of handling damaged material received from the client
- Method of controlling processes
- Corrective and preventive action procedure

Figure 5.32 Services checklist.

5.17 AUDIT PROCESS

5.17.1 The frequency for conducting a macro or micro audit is predicated on complexity of a product design, complexity and criticality of applicable product characteristics; past performance history; and the needs, interests, and expectations of both internal and external customers. Subsequent to the establishment of an appropriate audit checklist, an auditor will identify the verification stations where the checklist applies such as management, receiving inspection, in-process inspection, final inspection, gage calibration laboratory, and so on. The next action taken is the method of assessing each characteristic selected for evaluation. The number of observations made for each verification station and associated characteristic is usually selected via a sampling plan that is specified in a quality plan. An auditor might notice that an important audit factor representing an area of activity is important, but it is not covered in written procedures. When this occurs, the auditor will recommend that appropriate corrective action be taken.

5.17.2 Documented policy, procedures, and processes are reviewed prior to the application of an operation to determine if they will or will not fulfill requirements stated in a contractual agreement. When an auditor discovers a procedure to be inadequate or not available to operations personnel, the individual who retains the authority to approve the procedure will be immediately notified. Conditions like this should be avoided because objective contract management requires that quality procedures and processes be prepared in a timely manner, that they be all-inclusive, and that they be readily available when needed.

5.17.3 Corrective actions should be factual, objective, and related to active contract requirements. On-the-spot corrective actions may be authorized by an auditor when it is determined that a defect is minor and that follow-up action in not necessary. Deficiencies other than those considered minor should be documented and reported for necessary action. The report should show the date that the deficiency was detected and a reference number for identification purposes. A deficiency in a product may be classified as a variation or deviation to a specified requirement. A variation might be reworked or accepted as is. However, product characteristics in an as-is condition must be submitted to a client for approval. Product quality that deviates from a specified requirement is usually rejected. If it is determined that the deviation will not impact the end use of an item, a supplier may submit a request to its client for a contract change notice. A contract change is a written order signed by a contracting officer authorizing a request for a change in a contract. A request for a contract change usually triggers a request for a monetary consideration from a client. Approval of a variance (waiver) from the requirements of a drawing, specification, or other technical data referenced in a contract is made after the award of a contract. It is applicable only to specific contracts. For the purpose of technical evaluation, waivers pertaining to a product or service that does not conform to contract requirements are classified as a critical, major, or minor defect.

The director of quality assurance or a designated representative assume the responsibility for verifying that satisfactory and timely action is taken to rectify a reported complaint. In addition, auditors should include a reported nonconformity as a component of a subsequent audit.

NOTES

1. From ANSI/ISO/ASQ Q9001-2000: Quality management systems: Requirements, Clause 8.2.2.
2. From MIL-S-52779A: *Software Quality Assurance Program Requirements,* August 1, 1979.

6

Case Study 1: Processing an Invitation-for-Bid

6.1 CONTRACT

A contract that is based on the lowest price alone can be a false economy because it can lead unsatisfactory performance or late deliveries which result in added administrative costs to both customer and supplier.

6.2 SOLICITATION

6.2.1 On December 14, XXXX the XYZ Company, Anytown, U.S.A., distributed an invitation-for-bid to three companies for 25 each ball valves (see Figure 6.1). Two of them, Company Number One and Company Number Two, have a satisfactory quality history with the potential client regarding the production of similar valves. Company Three, also producers of ball valves, had no previous quality history with the XYZ Company. All of the solicited suppliers completed their bid package within an allotted time frame. It was subsequently determined that Company Number Three, Anytown, U.S.A., submitted the lowest bid. However, because its performance history was unknown, the XYZ Company elected to exercise an option that authorized them to verify the low bidder's quality assurance capabilities via an on-site audit before making a contractual commitment.

Solicitation Requirements
(An invitation-for-bid)

1. Issued by: XYZ Company, Anytown, U.S.A.

2. Bid number: 1002   3. Date Issued: December 14, XXXX

4. Name and address of bidder

5. Item no.	6. Supplies and services	7. Quantity	8. Unit cost	9. Total cost
1	Ball valve, for air, nitrogen, helium, or hydraulic service Specification no. VXXX100 Drawing number 1XXX23	25 ea.	_____	_____
1a	First article			NSP*
1b	Mechanical properties report			NSP
1c	Chemical analysis report			NSP

10. Special requirements:

     a. Policy manual in accordance with ANSI/ISO/ASQ Q9001-2000: Quality Management Systems: Requirements.

     b. Calibration system ISO 10012-2003: Measurement Management Systems: Requirements for measurement processes and measuring equipment.

     c. Quality Policy (QC manual), procedures, processes, and associated documentation shall be made available to the purchaser's quality assurance representative for review and acceptance.

     d. Inspection records shall indicate the nature of observations and the number of observations made.

     e. The supplier is responsible for the performance of all inspection and testing of supplies ordered. However, the customer reserves the right to perform any inspections and/or audits as deemed necessary to ensure that supplies and services conform to applicable drawings, specifications, and standards.

11. Packaging and marking: in accordance with standard commercial practice.

Figure 6.1   Solicitation requirements.   *(Continued)*

(Continued)

12. Inspection and acceptance:
    a. First article – source (at the supplier's plant)
    b. Production lot – source (at the supplier's plant)

13. Delivery performance:

Item	Quantity	Desired delivery after date of contract award
First article	1 each	90 days
Production	10 each	60 days
	14 each	120 days

NOTE: Enclose a copy of the mechanical properties and chemical analysis report of raw material with each shipment.

14. Special contract requirements:
    The quality of all products and services furnished shall conform to the highest, latest, state-of-the-art standards.

15. Post-award orientation conference
    To be convened ten days after the receipt of a contract award. Time and place will be announced under separate cover.

16. Discount for prompt payment

	Calendar days	Percentage
	10	_____
	20	_____
	30	_____

17. Sealed offers in original (no copies) for furnishing solicited supplies shall be received by the XYZ Company, Anytown, U.S.A., before 1:30 p.m., January 14, XXXX.

18. Name/title of the person authorized to sign the offer:

_____

19. Signature   20. Offer date

_____   _____

* Not separately priced

Figure 6.1 Solicitation requirements.

6.2.2 Company Number Three subsequently submitted its bid within the closing date referenced in the invitation-for-bid. (See Figure 6.2.)

Company Number Three
Anytown, U.S.A.

January 10, XXXX

XYZ Company
Anytown, U.S.A.

Subject: Response to invitation-for-bid
Reference: Bid number 1002

Bid is submitted as follows:

Solicitation Item	Bidder's offer
No. 8	Unit cost: $290.00
No. 9	Total cost: $7250.00
No. 12	Inspection and acceptance:

a. First article at Company Number Three, Anytown U.S.A.

b. Production lot at Company Number Three, Anytown, U.S.A.

No. 16   Discount for prompt payment:

Calendar days	Discount
10	7%
20	5%
30	3%
Over 30 days	None

No. 19   Signed/Chief Executive Officer

No. 20   Offer date: January 9, XXXX

Figure 6.2   Low bidder's response to the solicitation.

6.3 OVERVIEW

6.3.1 Prior to issuing a contract to a low bidder with unknown capabilities, a potential customer will audit the disciplines of financial, production, purchasing, and quality assurance capabilities. However, for this case study our primary focus is centered on the discipline of quality assurance.

6.3.2. It is recognized that a contract award is submitted to the lowest bidder. However, this may not always be cost effective in the long run. Therefore, the purchaser must be assured that a proposed supplier will deliver exactly what is purchased. Hence, the buyer's quality assurance specialist, armed with a copy of the invitation-for-bid and associated drawings and specifications, began to plan for the audit. This type of audit takes priority over routine contract administration, and it should be accomplished in a timely manner.

6.3.3 Empirical knowledge indicates that for this type of audit, seven working days are usually allocated to the audit cycle. The time cycle starts with a notice to a proposed supplier regarding the performance of an audit and ends with action taken the by the director of purchasing or a standing review board. An audit report must be free of ambiguity and differences of opinions, and it must address all aspects of quality requirements as well as the proposed supplier's quality history. Conclusions and recommendations are then made that support decisions made regarding a supplier's competence, capability, and responsibility.

6.3.4 When a capability requirement is a component of a solicitation, any exception by a low bidder that would allow the prospective client to verify capability will be considered as a negative response to a tender offer.

6.3.5 Conclusions and recommendations made in an audit report associated with a supplier's capability must be free from ambiguity and differences of opinions, and based on a proposed supplier's current and past experience that relate to a solicited item.

6.4 AUDIT PROCESS

6.4.1 On January 14, XXXX, the purchaser notified Company Number Three that John Q. McAssurance, its company's quality assurance specialist, will conduct an audit at his facility on February 4, XXXX. The purchaser's quality assurance specialist arrived at Company Number Three's facility on the morning of February 4, XXXX, as scheduled and had an introductory meeting with the plant manager and his technical advisors. The primary purpose of this meeting was to resolve any problems associated with the solicitation that might prevail in the proposed supplier's offer as well as to determine the name of the person empowered to speak for the company.

6.4.2 At the conclusion of this meeting, the plant manager advised the auditor that he and top management are fully supportive of the company's quality control department's quest for quality excellence. He suggested that his director of quality assurance provide the potential customer's quality assurance representative with a tour of the plant before beginning the audit and to give him a cursory review of the products that are in current production. He also was told to show the auditor how the company is currently applying its quality system. John Q. McAssurance subsequently began the audit by reviewing the adequacy of the company's quality assurance capabilities. This action was accomplished by examining the company's policies and procedures at three randomly selected product verification stations. His findings were then included in the following report.

6.5 AUDIT REPORT

Audit date: February 4, XXXX

Name of supplier: Company Number Three, Anytown, U.S.A.

Quality assurance organization:
    Director of quality assurance
    Quality engineer
    Calibration technician
    Chief inspector
    One receiving
    Two in-process inspectors
    One final inspector
    Customer complaint monitor

Employees contacted/years of experience
    Mark McQuality, Director of Quality Assurance (20 years)
    Sam Showthem, Quality Engineer (14 years)
    Martha Checkit, Chief Inspector (8 years)
    Cal A.Bration, Calibration Technician (11 years)

Ratio between inspection and production personnel 1:7

6.5.1 Evaluation Factors

	Adequate	Inadequate
Organizational structure	X	
Quality system	X	
Quality plan		X
Inspection and test records	X	
Inspection and test equipment	X	
Source of measurement standards	X	
Traceability of inspection and test records		X
Traceability of measuring instruments		X
Calibration procedures	X	
Adequacy of calibration procedures	X	
Record of calibrations	X	
Review of certificate of compliance		X
Objective quality evidence	X	
Reliability of inspection and test instructions	X	
Controls for investigation of customer complaints	X	
Identification of nonconforming products and services	X	
Documentation of nonconformity	X	
Documentation of corrective action taken	X	
Identification nonconforming cause	X	
Follow up of corrective action taken	X	
Timely issue of drawings and specifications	X	
Removal of obsolete documents	X	
Internal audits	X	
Preservation, packing, packaging, and marking	X	
Verification of certificates of conformance	X	
Method of identifying in-storage raw material	X	
Records	X	
Feedback data	X	
Control of specifications, drawings, change notices, modifications, and work/process instructions	X	
Control for selecting qualified suppliers	X	

6.6 OBSERVATIONS

a. The supplier is in the process of upgrading its quality system to meet the requirements of ANSI/ISO/ASQ Q9001-2000 Quality Management System: Requirements *(see Conclusions, Note 1).*

b. Records indicate that the supplier has a satisfactory performance record regarding the production of similar items.

c. The supplier understands the technical requirements of the bid.

d. Supplier's quality system does not include procedures for tracing product inspection and test characteristics to a specific contract *(see Conclusions Note 2).*

e. General purpose, inspection, measuring, and test equipment is not traceable to an appropriate measurement standard *(see Conclusions, Note 3).*

f. The company's quality plan for the solicited item was not addressed at this point and time *(see Conclusions, Note 4).*

g. The proposed supplier utilizes subcontractors from an established list of qualified suppliers. This list was established from suppliers who have achieved ANSI/ISO/ASQ Q9001-2000 certified status and other suppliers who have a favorable quality and delivery history with the company.

h. There is no evidence that feedback data associated with certificates of compliance are validated by the receiving inspection department *(see Conclusions, Note 5).*

i. The company's inspection procedures are focused on defect prevention.

j. Established procedures provide for the timely identification and correction of deficiencies.

k. The supplier has a program for the investigation and correction of internal and external customer complaints that is managed by a customer complaint monitor.

6.7 CONCLUSIONS

Note 1: The proposed supplier clearly understands specified quality system requirements and is in the process of updating its quality manual to meet solicited requirements.

Note 2: Supplier recognizes the need to establish procedures that cover the traceability of product inspection and test characteristics and advised that this condition will be rectified on or before February 14, XXXX. (See Figure 6.3, Major Milestone Schedule, Item 13.)

Note 3: The supplier is in the process of establishing a procedure for the traceability of inspection, measuring, and test equipment. (See Figure 6.3, Item 13.)

Note 4: The bid item is new to the company. However it does recognizes that quality planning will be addressed in a timely manner. (See Figure 6.3, Item 5.)

Note 5: The supplier provided assurance that the adequacy of certificates of compliance associated with mechanical properties and physical and chemical analysis of raw material will be validated by a representative of the quality control department, effective immediately. (See Figure 6.3, Item 11.)

6.8 RECOMMENDATIONS

6.8.1. A recommendation for an affirmative award is predicated on the following:

a. The plant manager and the director of quality assurance recognized the shortcomings of its quality system and provided convincing information that corrections will be accomplished within the time constraints of the contract. This commitment is supported with a Major Milestone Schedule. *(See Attachment A.)*

b. It was determined through the review of documents related to products previously produced that the company has a satisfactory performance record with its active customers. Records also indicate that the company has no open customer complaints.

c. It was also noted through conversation with key management personnel that the company has a positive attitude regarding objective quality management.

S/John Q. McAssurance, Auditor February 5, XXXX

Attachment A
Company Number Three, Anytown, U.S.A.

Re: Invitation-for-bid, number 1002 issued by the XYZ Company, Anytown, U.S.A.
Item: 25 each ball valves part number 1XXX23

Topic	Feb 1	Feb 14	Feb 21	Mar 6
1. Review solicitation	C			
2. Contract review (Subject to contract award)		S		
3. Upgrade quality manual	I	I	S	
4. Management commitment	C			
5. Quality plan		S		
6. Process control	C			
7. Calibration system			S	
8. Control of nonconforming products and services	C			
9. Correction and preventive action procedure	C			
10. Control of records	C			
11. Certificates of compliance		S		
12. Feedback data		S		
13. Traceability: Product & Measuring Instruments		S		
14. Internal audit			S	
15. Training	O	O	O	O

Slippage:

Item number	Cause of slippage	Action taken

Legend:
S = Scheduled completion date   I = In-process   C = Completed   O = Ongoing

Prepared by:
S/Sam Showthem, Quality engineer January 10, XXXX

Reviewed by:
S/I. M. Watching, Plant manager January 12, XXXX

Figure 6.3 Major milestone schedule.

6.9 EXIT INTERVIEW

During the exit interview, the auditor advised the supplier's plant manager that due to a strict policy of his company, auditors are not authorized to reveal their conclusions to potential suppliers because that is the domain of the purchasing manager. He further advised them that notification of the results of the audit will be provided immediately after his audit report is examined by his company's standing review board.

6.10 AUDIT REPORT

The audit report was subsequently signed by the auditor and delivered to a standing review board via the purchasing manager for adjudication. The board, consisting of department supervisors who represent purchasing, production, quality assurance departments, and a chairperson, reviewed the auditor's report and concurred with his recommendations. (See Figure 6.4.)

Proposed supplier: Company Number Three, Anytown U.S.A.

Product: Ball valve

Part number: 1XXX23   Specification number: VXXX100

Signature	Title	Decision	Date
S/Ron D. Rector	Director of quality assurance	Award	2-11-XX
S/I. Makeit	Production manager	Award	2-11-XX
S/Will Getit	Director of purchasing	Award	2-11-XX
S/Anna Rehview	Chair, standing review board	Award	2-11-XX

Figure 6.4   Decision of the standing review board.

6.11 NOTICE OF CONTRACT AWARD

On February 18, XXXX, the director of purchasing prepared a notice of the contract award. This notice (see Figure 6.5) was then attached to a copy of the solicitation and sent to Company Number Three.

From: XYZ Company, Anytown, U.S.A.

To: Company Number Three, Anytown, U.S.A.

Reference: Solicitation no. 1002

Contract no: C-1002

Item: Ball valve – 25 each drawing number 1XXX23

Total dollar value: $7250.00

Invoice: Submit to: Mr. Will Getit
                       Director of Purchasing
                       XYZ Company
                       Anytown, U.S.A.

Date of award: February 18, XXXX

S/Will Getit, Director of Purchasing, February 18, XXXX

Figure 6.5   Notice of contract award.

6.12 POST-AWARD ORIENTATION CONFERENCE

Shortly after issuing a signed copy of the contract, the customer's quality assurance representative began to formalize plans for a post-award orientation conference with the supplier's quality assurance manager. The primary purpose of the conference was to verify that promises made during the pre-award audit were fulfilled. The customer notified the supplier on February 20, XXXX, of his desire to conduct a post-award conference (see Figure 6.6) on February 24, XXXX. This date proved to be compatible with all participants.

Re: Request for a post-award conference

Date submitted: February 20, XXXX

From: John Q. McAssurance

Director of Quality Assurance

XYZ Company, Anytown, U.S.A.

To: Mark McQuality

Director of Quality Assurance

Company Number Three, Anytown, U.S.A.

Action item:

___   First-article inspection

___   Regular production inspection and testing

X   Post-award orientation conference

___   Resubmission inspection

Conference date: February 24, XXXX, at 9 am

Contract number: C-1002

Item description: Ball valve, drawing number: IXXX23

Quantity: 25 each

Figure 6.6   Memo.

6.13 CONFERENCE ATTENDEES

John Q. McAssurance, Director of Quality Assurance, XYZ Company

Mark Mc Quality, Director of Quality Assurance, Company Number Three

Sam Showthem, Quality Engineer, Company Number Three

Martha Checkit, Chief Inspector, Company Number Three

Cal A. Bration, Technician, XYZ Company

6.14 MAJOR AGENDA TOPICS

a. Traceability of product and related characteristics

b. Traceability of inspection, measuring, and test equipment

c. Quality plan

d. First-article requirements

6.15 TOPIC 1: TRACEABILITY OF PRODUCT CHARACTERISTICS

6.15.1 Observation

Provisions for tracing product characteristics are contained in the company's revised product/service observation record. This record provides the following information:

a. Product verification station

b. Contract/purchase order number

c. Inspection date

d. Product characteristic code number

e. The measuring device that was used to inspect a particular characteristic

f. Disposition (accept or reject)

g. Inspector who checked the product

h. Traceability of the product and the instrument that was used to inspect each product characteristic to a specific purchase order/contact

6.16 CORRECTIVE ACTION

Satisfactory corrective action taken. The supplier also advised that the observations will be included as an integral component of the company's quality audit checklists.

6.17 TOPIC 2

Traceability of general purpose inspection, measuring, and test equipment to standards of known accuracy.

6.17.1 Observation

a. Calibration procedures are now in place that provide an indication that inspection, measuring, and test equipment and measuring standards are traceable to standards of known accuracy which in turn are traceable to the National Institute of Standards and Technology.

b. Established calibration procedures identify the instrument's accuracy, its range, its discrimination, and the measurement standard that is used to calibrate the instrument. Procedures also provide for the traceability of product characteristics to the instrument that was used to check a product.

6.18 TOPIC 3: QUALITY PLAN

6.18.1 Observation

A copy of a contract as well as an updated and clearly defined abstract of contract quality requirements along with pertinent technical documents attached thereto, are the source documents that are used for preparing the quality plan. The following topics were reviewed and found to be acceptable:

a. Abstract of contract quality requirements

b. Summary of inspection and testing requirements

c. Responsibility of key managers quality assurance specialist

d. Selection of acceptable subcontractors

e. Method of evaluating subcontractor performance

f. Corrective and preventive action plan

g. Preliminary review monitor and material review board

h. Inspection and test status

i. Servicing

j. Training needs

k. The identification of pertinent drawings and specifications

l. The identification of product and service inspection and test characteristics

m. Traceability of product and measuring instruments

n. Required inspection and test procedures

o. Required inspection, measuring, and test equipment

p. Required measurement standards

q. Accuracy ratios between measuring instruments product tolerances

r. Accuracy ratio between secondary and general purpose measuring instruments

s. Accuracy ratio between primary and secondary measuring standards

t. Calibration procedures

u. Source of calibration procedures

v. Quality audit frequency

w. Method and frequency of reviewing quality assurance records

x. Method of controlling raw material

y. Method of preservation, packaging, packing, marking and shipping

6.19 TOPIC 4: CERTIFICATES OF CONFORMANCE

6.19.1 Observation

The supplier established a procedure that provides confidence that purchased raw materials used in the fabrication and processing of a product conform to specified physical and chemical analysis requirements. This is accomplished by reviewing the certificate of conformance against a specification and signed by the receiving inspector. In addition, the supplier's procedure stipulates that periodically a sample of raw material will be sent out to a qualified independent testing laboratory for verification analysis periodically.

6.20 FIRST ARTICLE

6.20.1 Observation

a. Performance of first-article inspection will take place within the supplier's facility on a test stand with pressure gages of known accuracy.

b. Instruments that will be used for pressure testing will be calibrated by an independent calibration laboratory with the use of a standard deadweight tester.

6.21 CONCLUSION

6.21.1 Quality assurance actions are performed at various stages of the manufacturing process to assure that the end item will conform to specified requirements.

6.21.2 No further action required. Satisfactory corrective action taken by the supplier to rectify all observed deficiencies.

7

Case Study 2: Application of Contract Quality Requirements

7.1 VERIFICATION OF QUALITY ASSURANCE CAPABILITIES

This case study is focused on the Wee-Cando Company. The company is a *jobber* that has no proprietary rights to any of the items that it produces for its clients.

The company's production, purchasing, subcontracting, and quality assurance functions are wholly dependent on technical documents and contract requirements that are provided by its customers. On January 29, XXXX, the Wee-Cando Company received a copy of an invitation-for-bid from the Wee-Needit Company, Anytown, U.S.A. to produce shaft assemblies. The Wee-Cando Company indicated to its potential client that it has a satisfactory performance history with its current customers. However, the Wee-Needit Company had no record of the proposed supplier's quality-assurance capabilities. The Wee-Cando Company responded to these concerns by inviting the potential customer to visit its facility and see first hand that they are capable of meeting solicited requirements. Agenda topics evaluated included:

a. Policy manual (Clause 7.2)

b. Communicating contract quality requirements (Clause 7.3)

c. Quality planning (Clause 7.4)

d. Inspection and testing (Clause 7.5)

e. Responsibilities (Clause 7.6)

f. Calibration of measuring instruments (Clause 7.7)

g. Correction and disposition of nonconformity (Clause 7.8)

h. Control of processes (Clause 7.9)

    i. Inspection and test status (Clause 7.10)

    j. Documents and data (Clause 7.11)

    k. Quality audits (Clause 7.12)

    l. Preservation, marking, packaging, storage, and shipping (Clause 7.13)

   m. Sampling techniques (Clause 7.14)

    n. Records (Clause 7.15)

    o. Customer service (Clause 7.16)

    p. Training (Clause 7.17)

    q. Staff meetings (Clause 7.18)

7.2 POLICY MANUAL

7.2.1 Wee-Cando records indicate that its policy manual was prepared by the company's quality engineer, reviewed by its quality control manager and approved by the company's plant manager. It also records the names of authorized recipients of its manual and maintains a record of past revisions. The policy manual was recently upgraded to meet the intent of ANSI/ISO/ASQ Q 9001-2000 Quality management systems: Requirements as it applies to current contractual obligations. Requirements that exceed those in the manual are established and implemented as specified in the customer's contract or purchase order. The information contained in the manual applies to in-house operations as well as product and service requirements delegated to vendors and subcontractors. The manual is reviewed and upgraded periodically in deference to continuous improvements in the quality system and to include proactive recommendations not referenced in the original manual. It is approved by top manager and distributed to recipients who have a need for the information contained therein. The quality assurance manager and other key management personnel report directly to the plant manager, who in turn reports to the president of the company. (See Figure 7.1.) Policy statements and sample documents referenced in the manual provide an indication to internal and external stakeholders as to how products and service quality is established, controlled, and maintained.

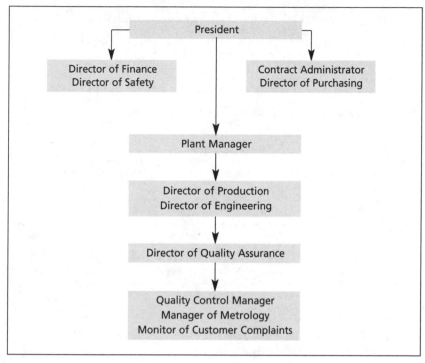

Figure 7.1   Organizational structure.

7.3 COMMUNICATING OF CONTRACT QUALITY REQUIREMENTS

7.3.1 The contract administrator or his/her designated representative is responsible for reviewing, summarizing contractual requirements, and documenting this information on an abstract of contract requirements. (See Figure 7.2.) A highlighted copy of the contract or purchase order is used to summarize noncomplex, noncritical items. Applicable documents are subsequently distributed to department supervisors who have a need of this information. An abstract of a contract is used as the source document for preparing each new quality plan. All questions or difference of options that might arise during the administration of contract requirements are directed to the contract administrator for resolution. The flow down of contract quality requirements to responsible department supervisors does not begin until there is assurance that contract requirements are clearly defined and understood.

Abstract of Contract Requirements – Customer	
Wee-Needit Company, Anytown, U.S.A.	Contract no: C - XXX-2000
Phone no: (1XX) XXX-1234	Fax: (1XX) XXX-2000
Name of purchaser's representative: Mr. John Doe	Date of Award: 1-22-XX

Discount terms: 2 percent, 30 days after receipt of contract award

1	First article
2	Shaft assembly—Drawing Number ABC-123

Point of product inspection and acceptance: First article (F/A) and production lots at Wee-Cando Company, Anytown, U.S.A.

Point of packaging inspection: Wee-Cando Company, Anytown, U.S.A.

Item no.	Quantity	Destination	Delivery dates
1 (F/A)	1 each	Wee-Needit Company, Anytown, U.S.A.	60 days
2 (Lot)	699 each	Wee-Needit Company, Anytown, U.S.A.	200 days

FOB Point: Wee-Needit Company, Anytown, U.S.A.   Drawing no: XXX-123

Specifications:

a. ANSI/SO/ASQ Q9001-2000 Quality management systems: Requirements

b. ISO 10012-1 Part 1: Metrological confirmation system for measuring equipment

c. ANSI/ASQ Z1.4 Sampling Procedures and Tables for Inspection by Attributes

Prepared by: S/A.D. Ministrator   **Date:** February 4, XXXX

Figure 7.2   Abstract of contract requirements.

7.4 QUALITY PLANNING

7.4.1 Planning is accomplished immediately after it is found that contract requirements are clearly defined and clearly understood, and that associated technical documents are readily available to management and operation personnel.

7.4.2 Factors addressed when establishing a quality plan include:

- Assurance that the quality plan, the abstract of contract requirements, and associated technical documents are distributed to respective members of a quality management team in sufficient time for the implementation of contract quality requirements

- Assurance that required technical capability to meet all specified contract quality requirements is acceptable

- Assurance that the company has, or is able to obtain, required resources including human, financial, production, and quality assurance capability within the time constraints referenced in a solicitation or contract

- The acquisition of additional personnel, plant facilities, and equipment, where applicable

- Required technical documents such as drawings, specifications, standards, and changes thereto are all-inclusive and clearly defined

- The need for a new policy, procedure, process, or work instructions

- Assurance that the quality plan is workable and that it provides confidence that all contract quality requirements will be met in an economical and objective manner

7.5 INSPECTION AND TESTING

7.5.1 Written instructions and test procedures are made available to management and operations personnel at each product verification station. The instructions include:

- Item nomenclature

- Part number

- Product characteristic

- Respective product characteristic

- Measuring device that is used to inspect the product

(See Summary of Inspection Requirements, Figure. 7.3.)

Item: Shaft Assembly	Part no.: ABC-XXX	
Characteristic code no.*	**Characteristic & tolerance**	**Measuring device**
101	Concentricity - datum axis A Tol. 0.003"	V-Block / Dial gages
102	Diameter 1.623" +/- 0.005"	Vernier caliper
103	Diameter 1.248" +/- 0.005"	Vernier caliper
104	Dimension 0.463" +/- 0.001"	Micrometer
105	Dimension 0.347" +/- 0.001"	Micrometer
106	Pin location 0.625" +/- 0.005"	Dial indicator
107	7/8-14 UNS-2A thread	Thread ring gage
201	Surface roughness 32 micro inch	Comparitor blocks
202	Material condition A 303 CRS	Certification

One hundred number series is the code assigned to product characteristics that are classified as major and the two hundred series are assigned to product characteristics that are minor in nature. They are usually an integral component of a drawing. When not shown on a drawing they should be identified in a related quality plan.

Prepared by:	Title:	Date:
John Q. N. Gineer	Quality Engineer	4-4-XXXX

Figure 7.3   Summary of inspection requirements.

7.5.2 "When contract quality requirements are established and made known to the quality control manager by the contract administrator, inspection and calibration can begin. One recommended method for communicating a contract quality requirement is to establish a master requirements list (MRL) (see Figure 7.4). The list is prepared for each product or assembly and requires modification only when there are significant changes to the product design or other contract quality requirements."[1]

MRL No. 5				
Part Name: Shaft Assembly	Part No.: ABC-XXX			
Chara. code no.	Chara. identification	Measuring device (MD)	MD ident. number	MD code number
101	Concentricity datumaxis A tolerance 0.003"	V-block/dial gage	20010	MD1
102	Diameter 1.623" +/- 0.005"	Vernier caliper	20011	MD2
103	Diameter 1.248" +/- 0.005"	Vernier caliper	20012	MD3
104	Dim. 0.463" +/- 0.001"	Micrometer	20013	MD4
105	Dim. 0.347" +/- 0.001"	Micrometer	20014	MD5
106	Pin loc. 0.625" +/- 0.005"	Dial indicator	20015	MD6
107	7/8-14 UNS 2A thread	Thread gage	20017	MD8
201	Surface roughness 32 micro inch	Comparator	20016	MD7
202	Material cond. A 303 CRS	Certification	Visual	Visual
Prepared by: John Q. N. Gineer	Title: Quality Engineer		Date: 4-4-XXXX	

Figure 7.4 Master requirements list.

7.5.3 Raw Material

Incoming raw material is accepted only when it is furnished with a certificate of compliance report. (See Figure 7.5.) The report is checked by the receiving inspector to assure that the data referenced therein is complete and that it is in compliance with specified requirements. When the report does not accompany the shipment, acceptance of the material is held in abeyance until one is received. Certified test reports that are accepted by incoming inspection authorizes the receiving inspector to release raw material from temporary to permanent storage. Questionable raw material will continue to remain in temporary storage until specific identity is accomplished. When certifications are determined to be unacceptable, the purchasing department will notify the vendor of this condition. (See Certificate of Compliance, Figure 7.6.)

Test Report No. x x x x x x x		

From: Supplier, Anytown, U.S.A.　　　　To: Customer, Anytown, U.S.A.　　　　4-12-XX

Product description: Type 303 stainless steel cold drawn rods

Specification: ASTM A 581/A 581 -95a Free-Machining Stainless Steel Wire and Wire Rods
Size: 0.375" round bar

Chemical Composition Characteristic	Specified values	Measured values
Carbon (max)	0.15	0.06
Manganese	2.00	1.79
Phosphorus (max)	0.20	0.04
Sulfur (max)	0.15 min	0.40
Silicon (max)	1.00	0.66
Chromium	17.0 -19.0	17.40
Nickel	8.0 - 10.0	8.67
Mechanical Test Requirements	**Specified values**	**Measured values**
Yield strength: PSI	115 to 145	590
Tensile strength: PSI	795 to 1000	772
Elongation	1.25" %	1.0%

Signature: I. Jane DeSupervisor　　　　　　　　　　　　　3-20-XX

Disposition: Report: Acceptable _X_　　　Not acceptable ____

S/John N. Spector　　　　　　　Receiving inspector　　　　4-18-XX

Figure 7.5　Raw material certificate of compliance.

Vendor:
Provider of Raw Material
Anytown, U.S.A.

Date: May 15, XXXX

Re: Certificate of Compliance

Item: Condition A 303 CRS

Job no.: 021   Quantity: 3000 pounds   Purchase order no.: 000456

Dear Sir/Madam,

Material on the purchase order referenced above has been received without the accompanying certificate of compliance. Without this certificate, we are unable to process your material through receiving inspection. The material will be held in abeyance until the certificate is received.

Sincerely,

John I. DiRector, Purchasing Department

Figure 7.6   Request for certificate of compliance.

7.5.4 Customer-supplied Product

7.5.4.1 Customer-supplied product (CSP) is inspected upon receipt for count, condition, and quality status. Inspection of the product is accomplished within five days after receipt of CSP. When a shipping container is received in a damaged condition, a determination is made as to its impact on the quality status of its contents. When the product is found to be damaged, the package and dunnage is discarded after satisfactory corrective action is taken. Shipping discrepancies are defined as:

  a. Shortages or overages within a container

  b. Incorrect part number or stock number

  c. Misdirected shipments

  d. Missing or incorrect paper work

  e. Conditions detected at the time of receipt that materially affects usability

Shipments received with visual damage are processed as follows:

a. Annotate carrier's receipt describing observed damage to the product

b. Notify the carrier immediately via telephone and confirm by fax

c. Secure exhibit and where appropriate photograph the damage

d. Request an inspection from the carrier's representative

e. Immediately notify the customer's representative and provide him/her with a copy of all documents

Items are identified as follows:

a. Tagged or marked as CSP

b. Include details:
   1. Item identification
   2. Contract or purchase order number
   3. Item serial number
   4. Control number
   5. Applicable job number

7.5.5 Item identification should be permanent, legible, tamper-proof, and placed conspicuously on the product or container. Accepted CSP is segregated from company property and it is stored in a designated holding area. A representative of the quality control department is then assigned the responsibility for the protection, accountability, and periodic inspection of all CSP. The use of CSP for any purpose other than specified in a purchase order or contract is forbidden unless expressly authorized in writing from the customer.

7.5.6 Final as well as receiving and in-process inspection is performed in accordance with an established quality plan. Associated records are filed in the respective job order file and they are readily available to both internal and external auditors. Completed inspection and test records are placed in storage where they are protected from damage, deterioration, or loss. They remain there for a minimum of three years or as otherwise specified in a contract or purchase order.

7.5.7 Attribute and variable-type product characteristics are documented. (See Figures 7.7 and 7.8.) Accepted production lots are identified with an Acceptable Product Tag and routed to its next destination. Rejected lots are identified with a red rejected product tag and routed to a designated holding area for necessary corrective action and disposition. The rejected product is resubmitted into the production cycle only after sorting, re-inspection, and purging of nonconforming products. Traceability of a product characteristic and the instrument used to inspect a product is accomplished as follows:

a. The metrology department assures that all measuring standards and inspection, measuring, and test equipment are calibrated with instruments of higher accuracy levels that are traceable by an unbroken chain of calibration events to the National Institute of Standards and Technology.

b. Product characteristics are traceable to a specific measuring instrument and an applicable contract or purchase order number.

c. End pieces of bar stock are color-coded, and they are traceable to a specific heat number and purchase order.

Shaft Assembly Part Number ABC-123			
Date	Sample size	Number of defectives	Percent defective
May 1	50	5	.10
2	50	8	.16
3	50	12	.24
4	50	17	.34
5	50	15	.30
8	50	4	.08
9	50	12	.24
10	50	8	.16
11	50	15	.30
12	50	20	.40
15	50	10	.20
16	50	9	.18
17	50	8	.16
18	50	6	.12

Figure 7.7 Documentation of attribute type product characteristics. *(Continued)*

(Continued)

Shaft Assembly Part Number ABC-123			
Date	Sample size	Number of defectives	Percent defective
May 19	50	8	.16
21	50	23	.46
22	50	7	.14
23	50	10	.20
24	50	5	.10
25	50	25	.50
June 1	50	9	.18
2	50	10	.20
3	50	12	.24
4	50	2	.04
5	50	13	.26

Total number of observations = 1250

Number of defectives = 273

Percent defective = 0.22

Figure 7.7 Documentation of attribute type product characteristics.

Contract no: C-XXX-2000						
Master requirements list: No. 5			Inspection station: No. 2			
First article: _____	Receiving: _____		In-Process: _____		Final: _____	
Part nomenclature: Shaft Assembly		Drawing no: ABC-XXX			Quantity: 1 ea.	

Chara. code no.	Drawing dimension	Measuring device	Observed obs. value	Disp.	Inspector	Date
101	Concentricity w/datum Axis A Tol. 0.003"	V-block/dial gage	0.001"	Ac	MRP	6.1 XX
102	1.623" +/- 0.005"	Vernier caliper	1.624"	Ac	MRP	6-1-XX
103	1.248" +/- 0.005"	Vernier caliper	1.248"	Ac	MRP	6-1-XX
104	0.463" +/- 0.001"	Micrometer	0.461"	Ac	MRP	6-1-XX
105	0.347" +/- 0.001"	Micrometer	0.347"	Ac	MRP	6-1-XX
106	Pin loc. 0.625 +/- 0.005"	Dial indicator	0.627"	Ac	MRP	6-1-XX
107	7/8-14 UNS 2A thread	Thread gage	Visual	Ac	MRP	6-1-XX
201	Surface 32 micro in. RMS	Comparator	Visual	Ac	MRP	6-1-XX
201	Material cond. A 303CRS	Certification	Visual	Ac	MRP	6-1-XX

Number nonconforming: None

Reviewed by: Quincy C. McManager   Title: Quality Control Manager   Date: 6-2-XX

Audited by: Ida McAdider   Title: Audit Specialist   Date: 6-3-XX

Figure 7.8   Record of variable-type product characteristics.

Note: When a contract specifies a requirement for concurrent inspection or testing at a supplier's plant, the client is notified via a written notice of production lots that are ready for concurrent inspection and testing. (See Figure 7.9.)

Source Inspection		
Date requested: April 20, XXXX		
From: Director of Quality Assurance Wee-Cando Company Anytown, U.S.A.		
To: Director of Quality Assurance Wee-Needit Company Anytown, U.S.A.		
Part nomenclature:	**Drawing no:**	**Specification no:**
Shaft assembly	XXX123	N/A
Scheduled inspection/testing date: April 27, XXXX		
Comments: a. The product will be ready for *concurrent inspection, which is scheduled for 9 a.m. in the final inspection department*		
Signed: John Q. McAssurance	Director of Quality Assurance	4-20-XX

Figure 7.9   Request for concurrent inspection.

7.6 RESPONSIBILITIES

7.6.1 Contract Administration

The contract administration department is the focal point regarding all actions associated with contract compliance. They maintain a control file for each contract. The file includes but is not limited to:

a. Copy of a solicitation

b. Copy of the contract

c. Correspondence

d. Copies of certifications

e. Shipping documents

f. Invoices

g. Record of customer complaints

7.6.2 Quality Control Department

a. Interprets contract quality requirements

b. Identifies the type and amount of inspection that will be required to meet contract quality requirements

c. Reviews technical documents

d. Prepares required inspection testing instructions

e. Plans, develops, initiates, and maintains, with top management's support, the most effective and efficient quality system for optimum quality assurance performance

f. Maintains objective quality records

g. Conducts quality audits

h. Reviews the adequacy of corrective actions taken by internal and external suppliers

i. Monitors vendor and subcontractor quality performance

j. Assures that inspection personnel are capable of rendering unbiased decisions regarding the acceptance (or rejection) of products or processes inspected / evaluated by them

7.6.3 Purchasing Department

The purchasing department is responsible for that part of quality systems management concerned with the procurement of supplies and services required for the support of manufactured products. Activities of the purchasing department include, but are not limited to, the following:

a. Coordinates the selection of qualified suppliers of products and services

b. Alerts subcontractors of inspection and test requirements

c. Notifies vendors and subcontractors of product and / or service deficiencies observed by the quality control department

d. With support from company quality assurance personnel, assures that subcontracted products and services conform to specified requirements

e. Furnishes the quality control department with test results and certificates of conformance received from subcontractors

f. Specifies source inspection at subcontractor facilities where it is not practical to inspect a delivered product at its delivery point

g. Supplies required specifications and drawings to sub-suppliers regarding contract requirements delegated to them

h. Procures products and services from suppliers who have a record of providing acceptable products and services

7.6.3.1 *Selection of Qualified Suppliers.* Suppliers are separated into two categories:

- Those that produce complex items and services

- Those that produce noncomplex items and services

7.6.3.2 For this case study focus is centered on those suppliers who are involved with the production of higher-level, complex items. A list of higher-level items is prepared in advance of initial production. The purpose of this list is to provide quality assurance personnel and the director of purchasing with advance notice regarding the need for special planning. This action is important for items that that are complex in nature. There are several reasons for listing the purchase of complex items. This information is needed to determine:

a. Internal as well as external quality assurance capabilities

b. The degree of in-process controls to be implemented

c. The amount of final inspection that will be required

d. Site of in-process and final inspection

e. Concurrent product inspection and testing (where appropriate)

f. The establishment of an audit schedule

g. The extent of quality control requirements delegated to an outside supplier of products and services

h. To support the purchasing department in determining if there is a need to outsource a part or all of CQR

7.6.3.3 Purchasing, with support from the quality control department, notifies its suppliers of observed noncompliance with contractual quality requirements, and the quality control manager or his/her designated representative performs follow-up action to assure that satisfactory corrective action is taken.

7.6.3.4 *Rating the Supplier:* A supplier rating system is used to evaluate the quality level of products and services furnished by outside sources. It is applied as follows:

a. Purchasing, in concert with the quality control department, is responsible for the use and control of this system. Incoming inspection specifically plays an important role in applying the system. The receiving inspector informs the purchasing department via normal channels regarding a supplier's ability to continually furnished acceptable products and/or services.

b. Suppliers and vendors are continually evaluated. (See Supplier Rating Record Figures 7.10 and 7.11.) The record includes the lot number, vendors name, date, purchase order number, number of lots inspected, number of lots rejected, evaluation date, disposition, name of the receiving inspector and quality control manager, and appropriate comments.

7.6.3.5 The supplier's quality performance rating is achieved by dividing the *number of lots accepted,* by the *number of accepted lots inspected* and then multiplying by 100. This will give the rating in percentage. For example:

$$\frac{\text{Number of accepted lots} = 8}{\text{Number of lots inspected} = 10} \times 100 = 80$$

Suppliers are rated as follows:

- A rating of 98 percent or greater is considered acceptable (Good).

- A rating of 90 percent to 97 percent is a negative trend (NT). No further purchases will be made from this vendor until satisfactory corrective action is taken.

- A rating of 89 percent or less is considered unacceptable; seek another source of supply (SAS) when satisfactory corrective action is not taken.

7.6.3.6 All suppliers with a satisfactory history and well as those with an unsatisfactory quality history is summarized by the quality control department on a Summary of Supplier Performance Form. (See Figure 7.12.) A copy of this report is then furnished to the purchasing department for appropriate action.

Acceptable Supplier					
Vendor: Company A, Anytown, U.S.A.					
Item: Various					
Lot no.	Purchase order no.	Date	Disposition	Inspector	Rating
1	D330012	1-4-XX	Accept (Ac)	MRP	
2	D330016	3-1-XX	Ac	MRP	
3	D330017	3-8-XX	Ac	MRP	
4	D330022	4-27-XX	Ac	MRP	
5	D330026	5-4-XX	Ac	MRP	Good
6	D 330030	5-11-XX	Ac	MRP	
7	D330037	5-21-XX	Ac	MRP	
8	D330039	5-23-XX	Ac	MRP	
9	D330046	9-1-XX	Ac	MRP	
10	D330050	11-5-XX	Ac	MRP	Good
Comments:	a. Frequency of evaluation: Every 5 lots				
	b. Total number of lots inspected: 10				
	c. Total number of lots accepted: 10				
	d. Number of lots rejected: None				
	e. Evaluation dates: See Rating column				
S/Quincy C. McManager	Quality Control Manager	11-5-XX			

Figure 7.10   Supplier rating record – acceptable supplier.

Unacceptable Supplier					
Vendor: Company B, Anytown, U.S.A.					
Item: Various					
Lot no.	Purchase order no.	Date	Disposition	Inspector	Rating
1	D30011	2-4-XX	Accept (Ac)	MRP	
2	D330014	2-1-XX	Ac	MRP	
3	D330015	2-30-XX	Ac	MRP	
4	D330040	6-1-XX	Ac	MRP	
5	D3300443	8-2-XX	Ac	MRP	Good
6	D330045	8-30-XX	Ac	MRP	
7	D330048	9-29-XX	Ac	MRP	
8	D330049	10-12-XX	Ac	MRP	
9	D330051	11-8-XX	Reject (Re)	MRP	NT
10	D330070	11-11-XX	Re	MRP	SAS
Comments:	a. Frequency of rating: Every 5 lots				
	b. Total number of lots inspected: 10				
	c. Total number of lots rejected: 2				
	d. Evaluation dates: See "Rating" column				
S/Quincy C. McManager	Quality Control Manager	11-11-XX			

Figure 7.11   Supplier rating record – unacceptable supplier.

Distribution:	a. Purchasing
	b. Plant Manager
	c. Quality Control

Evaluation period

From: January 4, XXXX

To: November 11, XXXX

Supplier	No. of lots inspected	No. of lots accepted	No. of lots rejected	Rating
Company A Anytown, U.S.A.	10	10	None	100%
Company B Anytown, U.S.A.	10	8	2	80%*

*Conclusion

1. Company B is nonresponsive regarding requests for corrective action submitted to that organization by the quality control department. The cited deficiencies relate to purchase order numbers D330051 and D330070.

2. It is recommended that future purchases with this company remain in abeyance until satisfactory corrective action is taken.

S/Quincy C. McManager   Quality Control Manager   11-11-XX

Figure 7.12   Summary of supplier performance.

7.7 CALIBRATION OF MEASURING INSTRUMENTS

7.7.1 Calibration policies and procedures are prepared accordance with the intent of International Standard, ISO 10012-1:1992: Quality Assurance Requirements for Measuring Equipment – Part 1: Metrology Confirmation System for Measuring Equipment.

7.7.2 The flow down of contract quality requirements along with a quality plan is furnished to the metrology department in ample time so that specified technical requirements can be reviewed and analyzed prior to establishing/upgrading the calibration system.

7.7.3 Calibration procedures are normally prepared and provided by the manufacturer of an instrument. When they are not available from the instrument's manufacturer, they are prepared by the Wee-Cando Company or by an independent laboratory.

7.7.4 Calibration Status

The calibration status of each instrument is identified. This identification includes instruments that are calibrated *prior-to-use, general use, limited use,* and *do not use* status. (See Figure 7.13.) Instruments that are awaiting calibration or repair are separated from those that are usable and under the control of the metrology department.

Title	Application	Form no.
Prior to use	Each time the instrument is used	P2000
General use	Scheduled interval	G2001
Limited range	Limited use	L2002
Rejected Label	Do not use	R2003

Figure 7.13   Calibration status.

7.7.5 Calibration Sources

Calibration services are obtained from independent calibration laboratories who have verified capabilities that comply with the intent of ISO 10012-1-1992: Quality assurance requirements for measuring equipment – Part I: Metrological confirmation system for measuring equipment.

7.7.6 Certificate of Conformance

Certificates of conformance associated with calibration services provided by an outside source are traceable by an unbroken chain of events to the National Institute of Standards and Technology.

7.8 CORRECTIVE ACTION AND DISPOSITION OF NONCONFORMITY

The path to the identification, segregation, disposition, and correction of the cause of nonconforming product and service involves factors. (See Figure 7.14.) The corrective action program is established and maintained by the quality control department with support for a preliminary review monitor, a material review board, and a customer complaint monitor. All nonconforming products are initially identified on a red tag and routed to a designated holding area. The information entered on this tag includes the applicable purchase order or contract number, date, part number, quantity, responsible department, a brief description of the nonconformance, and the inspector's signature. (See Figure 7.15.) The reported nonconformance will then be brought to the attention of the preliminary review monitor for adjudication within his/her assigned authority. He or she will determine if appropriate corrective action can be taken that is within the confines of a contractual agreement. If it is determined that a decision is beyond his or her authority, the reported nonconformance will be processed through the material review board and customer complaint monitor.

7.8.1 Corrective Action Process

- Attach nonconforming tag to production lot

- Route nonconforming production lot to a holding area

- Preliminary review

- Material Review Board

- Scrap

- Repair to standard procedures

- Rework to specified requirements

- Accept as-is with customer approval

- Notify department supervisor

- Notify subcontractor where appropriate

- Identify cause

- Correct

- Screen

- Re-inspect

- Document

- Review, analyze, and take appropriate follow-up action

Figure 7.14 Correction and disposition of nonconforming products and services.

Part name: Shaft	Department: Final inspection
Contract no: C-XXX-2000	Lot size: 700 each
Part no: XXX123	Job no: 1234

Reason for rejection:

a. Twelve each, dimension 0.463" plus/minus 0.001" is 0.001" undersize.
   (Ref. code number 104)*

b. Eight each surface roughness exceeds the allowable limit of 32 micro finish.
   (Ref. code number 201.)

* See Figure 7.8.

Inspector: S/Jane Doe III	Final Inspection Department 6-12-XXXX

Figure 7.15   Rejected product tag.

7.8.2 Preliminary Review

Preliminary review action (see Figure 7.16) is the responsibility of the quality control manager. He determines if the reported nonconformance can be eliminated by rework, if the product is to be scrapped or returned to the vendor because it is unfit for use, or whether the complaint should be referred to the Material Review Board (MRB).

7.8.3 Material Review Board Action

A MRB consists of representative of the quality control department, a representative from the engineering department and, where appropriate, a customer's designated representative. They meet when called upon to determine whether a product is to be reworked, accepted as is, returned to a vendor or scrapped. The quality control department has the responsibility for maintaining a record of all actions taken by the board, records of previous acceptance or rejections of similar items on a contract, and follow-up on action taken to preclude future occurrences of the same or of a similar nature. Items that require MRB action are placed in a designated holding area with the MRB report form attached to it. The decision of the board is final. A reject recommendation that is made by any member of the board is final. When acceptance is made, each member signs the MRB Form. (See Figure 7.17.) An item that is to be re-worked is then routed to the production department with the MRB report and rework order attached. After rework, the item is re-inspected and processed through normal production channels. Rejected items are routed to a holding area until they can be returned to a vendor or scrapped.

7.8.4 Customer complaints

When a justified complaint is received from a customer, the quality assurance manager or his designated representative will address the applicable factor referenced below:

    a. Examine records for completeness and for an indication of the quality of the product at the time that is was shipped to the customer.

    b. Where appropriate, request an exhibit of the nonconforming product.

    c. Determine the condition of the shipping container at the time the product was accepted by the customer.

    d. Determine the quality status of similar items in stock or current production.

    e. Where appropriate, examine the nonconforming product and shipping containers in the presence of the customer.

    f. Furnish a report of cause identification and corrective action taken to the complaining activity immediately upon completion of an investigation.

    g. Address products, procedures and processes impacted by a report of nonconformance in each scheduled audit until it is determined that satisfactory corrective action has been taken.

7.8.5 When it is determined that an extended period of time will be required to rectify an internal or external customer complaint, a major milestone schedule is prepared. Progress reports of action taken are sent to the originator of the reported nonconformity.

Part name: Shaft assembly	Drawing no. XXX-123	Lot size: 700 each
Part no.: XXX-123	Specification no.: N/A	Contract no.: C-XXX-2000
Number accepted: 680	Number rejected: 20	Job no.: 2001

Nature of defect:

a. Twelve each, dimension 0.463" plus/minus 0.001" is 0.001" undersize (characteristic code number 104)

b. Eight each observed with surfaces that exceed 32 micro finish (characteristic code number 201)

Cause of the defect: Worn cutting tools in both cases

Disposition:

a. Screen the lot

b. Rework surfaced to comply with specified 32 micro finish

c. Refer the undersize dimension condition to the material review board for disposition

Corrective action: (See MRB report, Figure 7.17 dated 6-17-XXXX)

Distribution

a. Quality control

b. Quality engineer

c. Plant manager

d. Material Review Board

e. Attach copy to the production lot

Signature: I. M. Monitor   **Date:** 6-15-XX

Figure 7.16   Preliminary review – nonconforming product.

Part Name: Shaft Assembly	Part no. XXX123	Date: 6-17-XX
Job no.: 1234	Quantity withheld: 12 ea.	Department: Machine shop
Contract no.: C-XXX-2000	Withheld by: Inspector Jane Doe I	

Reason for withholding	Frequency	Material Review Board action
Dimension 0.463" is 0.001" undersize (Twelve each)	First time	a. The undersized dimension does not effect form, fit, or function. Permission granted from the customer to accept the product as-is for a consideration of $100.
		b. Ship the parts to the customer as is.
		c. Attach a copy of this report to the shipping document.

Signatures (MRB)	Department	Date
S/John Q. Engineer	Engineering	6-18-XX
S/John Q. Control	Quality control	6-18-XX
S/George Madit	Production	6-18-XX
S/I. L. Buyit	Customer's representative	6-18-XX

Assignment of recommended action: manufacturing department

a. Replace worn cutting tool

b. Increase surveillance inspection

Reviewed by: Will Watchall, Plant Manager	Date: 6-19-XX

Figure 7.17 Material review report.

7.9 CONTROL OF PROCESSES

First piece inspection is accomplished after each machine setup. Inspections are also performed whenever a change is made in a manufacturing setup. Periodically thereafter, inspections are performed to assure that the processes used to create the product are continually maintained.

7.9.1 Product verification stations are established at various locations within the plant to facilitate the examination of product quality during fabrication and processing. (See content of floor plan, Figure 7.22.) Inspections are performed on all products received from production lines using drawings, specifications, and checklists.

7.9.2 In carrying out their assigned functions, the quality control department is guided by instructions referenced in an applicable quality plan and job order folder.

7.9.3 Statistical process controls are also implemented for the purpose of upgrading the processes that create the product.

7.9.4 A shop routing record accompanies all production lots throughout the production cycle. This document shows the operations, processes, inspection, and tests performed by the purchasing, production, and inspection departments. (See Figure 7.18.)

Part name:	Drawing no:	Revision no:	Job order no.
Shaft assembly	XXX-123	None	2001

Contract no:	Lot size:	Prepared by:	
C- XXX-2000	700 each	John. B. McEngineer, Project Engineer	

Oper. No.	Operation	Duration	Hours	Date	Employee	Ac	Re.
1	Order bar stock (raw material)			3-07-XX	*		
2	Receive/inspect raw material	10-11 a.m.	1	4-16-XX	A1	All	
3	Code raw material	2-3 p.m.	2	4-16-XX	**		
4	Machine first article	8 a.m.-4 p.m.	8	6-03-XX	**		
5	Inspect first article	9 a.m.-1 p.m.	4	6-05-XX	A2	1	0
6	Produce production lot	8 a.m.-4 p.m.	8	6-11-XX	**		
7	Inspect production lot	8-10 a.m.	2	6-12-XX	A2	660	20
8	Preserve the product	9-10 a.m.	1	6-14-XX	**		
9	Inspect preservation	10-11 a.m.	1	6-14-XX	A2	660	0
10	Package and mark containers	11 am-1 p.m.	2	6-14-XX			
11	Inspect packaging and marking	1-2 p.m.	1	6-14-XX	A3		
12	Prepare shipping documents	8-9 a.m.	1	6-15-XX			
13	Notify carrier	10 a.m.	—	6-15-XX			
14	Ship the product	9 p.m.	—	6-16-XX	A3	700**	
15	Prepare/send billing	2 p.m.	—	6-16-XX	*		

*Purchasing department   ** Manufacturing.
Note: MRB disposition of nonconforming product as follows:

a. Twelve each dimension is 0.001 inch undersize accepted as condition

b. Eight each surface roughness reworked to comply with specified requirements

c. Twenty each of items (a) and (b) above reworked, reinspected, and included in the shipping

Reviewed by: I. M. McManager	Approved by: Will McWatchet
Quality Control Manager	Plant Manager

Figure 7.18 Shop routing record.

7.10 INDICATION OF INSPECTION STATUS

The issue, use, and control of inspection stamps that are used to indicate the inspection status is the responsibility of the quality control manager. The quality control manager indoctrinates inspection and test personnel regarding the proper use and application of each stamp. (See Figure 7.19 for stamp configuration.)

Figure 7.19 Configuration of inspection stamp.

7.10.1 A record of all stamps issued, which includes active as well as inactive stamps, is maintained on an Inventory of Inspection Stamps (Figure 7.20) and on an individual stamp distribution record. (See Figure 7.21.) Stamp recipients submit a written report to the quality control manager regarding the loss of any inspection stamps assigned to them. Subsequent to the receipt of this report, the quality control manager will prepare and distribute a memo that identifies the lost stamps to all affected department supervisors. Stamps returned because of employee transfer or termination will not to be reissued until after 24 months have lapsed. Stamps that have been lost will not be reused until 24 months after the date of the reported loss. Inspection stamps are used to identify the inspection and test status of products and services that are processed through the various inspection stations shown on a floor plan.

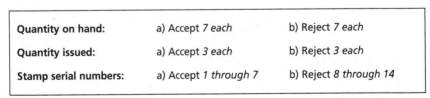

Figure 7.20 Inventory of inspection stamps.

Stamp type	Serial no.	Recipient	Date issued	Date returned
Accept	A1	Jane McGoodstamper	4-8-XX	
Reject	R8	Jane McGoodstamper	4-8-XX	
Accept	A2	Jim McCheckit	4-8-XX	
Reject	R9	Jim McCheckit	4-8-XX	
Accept	A3	John Wilcheck	4-8-XX	
Reject	R10	John Wilcheck	4-8-XX	

Comments:

Issued by:
Quincy C. McManager, Quality Control Manager

Figure 7.21 Distribution of inspection stamps.

Department	Action
Offices	Administrative
Conference room	Administrative
Sales office	Administrative
Quality control	Administrative
Manufacturing	Production
Processes	Verification station no. 1
Assembly	Verification station no. 2
Metrology	Verification station no. 3
Receiving inspection	Verification station no. 4
In-process inspection	Verification station no. 5
Final inspection	Verification station no. 6
Storage	Verification station no. 7
Packaging and shipping	Verification station no. 8

Figure 7.22 Content of floor plan.

7.11 DOCUMENTS AND DATA

Procedures are established, implemented, and maintained to assure that the latest drawings, specifications, and contract change notices are available when needed. It is the responsibility of the engineering department to release and control all parts lists, drawings, bills of material, operation sheets, and specifications used in the manufacture of parts and assemblies. A master control file of drawings and specifications is maintained under the control of the engineering department. A record of all design changes and modifications is maintained. All new or revised drawings and engineering changes are issued to production and the inspection departments through the quality control department. Obsolete drawings are immediately removed from all active files and marked "obsolete." Original drawings are immediately replaced when they become illegible. The current issue of drawings, specifications, and standards and associated policy and procedures, are traceable to open contracts and an applicable quality plan.

7.12 QUALITY AUDITS

When internal and external quality audits are implemented, each are carefully considered when preparing a quality plan. It is recognized that the bulk of scheduled audits are centered on the internal functions of quality system management, however external suppliers are not ignored. Unless otherwise specified, audit focus is centered on quality elements that are traceable to a specific contract (or invitation-for-bid) and associated policies, procedures, and processes.

7.12.1 Internal quality audits are performed to verify compliance with all aspects of the adopted quality system standard as well as meeting the clients needs and expectations. Quality audits may be performed by an independent quality assurance consultant or by someone selected by the quality control manager who does not have direct responsibility for the activity being audited. Established policies and procedures, along with the contents of a quality plan, are addressed when establishing quality audit checklists. External (subcontractor) audits are performed to the degree necessary to assure that quality requirements delegated to them are properly implemented and maintained.

7.12.2 A written report is prepared for each quality audit. The completed report, with a copy of the applicable checklist attached thereto, is submitted to the quality control manager with an information copy to the plant manager. The report addresses the following:

a. Product verification stations audited

b. Quality system elements audited

c. Policy procedures audited

d. Operation procedures audited

e. The identification of systemic factors audited

f. The identification of measuring instruments and associated measurements audited

g. Conclusions and recommendations

h. Name, title, and date of the person(s) who conducted the audit

i. Name and title of the person who reviewed and approved the report as well as the date it was reviewed and approved

7.12.3 Immediate action is taken to rectify reported deficiencies and/or recommendations made by the auditor. Where an extended period of time is required to rectify a reported complaint, an anticipated cure date will be provided to the requesting activity by the department supervisor assigned the responsibility for taking necessary action. Under this condition, periodic progress reports are furnished to the quality control department in advance of a final report.

7.13 PRESERVATION, MARKING, PACKAGING, STORAGE, AND SHIPPING

Products that are routed to the shipping department and storage are inspected to assure that the product is not damaged due to improper handling and storage. The quality control department conducts surveillance inspections in the shipping and storage areas to assure that items presented for shipment are processed in accordance with specified requirements. This includes the preservation, packing, packaging, and container marking and associated shipping documents. When packaging instructions are not identified, shipping containers will be processed in accordance with best commercial practices. Items will be shipped in a manner that will withstand physical and mechanical damage to the shipping container and its contents. Reports of shipping damage are thoroughly investigated by the quality control department to determine the cause of the reported complaint and required corrective action.

7.14 SAMPLING TECHNIQUES

Attribute type sampling procedures are established and maintained in accordance with those sampling methods and procedures specified in a customer's contract or purchase order. When a sampling plan is not specified in an invitation-for-bid, a contract, or a purchase order, the plan is established in-house. Variable-type characteristics are implemented for the purpose measuring the capability of a process, to identify methods that will improve a process and to identify ways to reduce costs by reducing variation of a selected characteristic that is most representative of a process.

7.15 RECORDS

Records are reviewed during internal audits to assure continued compliance with the latest contract quality requirements and associated procedures. They are established and maintained for work to be accomplished as well as work already accomplished. Records referenced throughout the policy manual are audited periodically to assure conformance with established company policy and procedures. They are filed in a manner that protects them from damage, deterioration, or loss. They remain in storage for a minimum of three years or as contractually specified by the customer.

7.16 CUSTOMER SERVICE

Engineering, production, and quality control departments along with a customer complaint monitor are delegated with the responsibility for assuring that services furnished to clients do in fact conform to purchase agreements. Internal and external customer complaints are continually evaluated. Responsible managers assure that immediate and satisfactory corrective action is taken to rectify justified client complaints. They assure that the causes of nonconformity are properly identified and that satisfactory corrective action will be taken to preclude a recurrence of service complaints.

7.16.1 Employees are instructed that satisfactory customer service is achieved through:

a. Clear communication

b. Understanding the client's needs and expectations

c. Prompt reaction to customer needs and expectations

d. Delivery of a services on time

e. Prompt and friendly response to client complaints

7.17 TRAINING

The following ten factors are addressed when planning, identifying, developing, and upgrading the level of expertise of each employee:

7.17.1 Training Needs

Identify the current level of employee education and experience. This is accomplished by comparing the education and training experience provided by employees on their application for employment and compared with training needs and objectives.

7.17.2 Training Requirements

Select training requirements for employee from an established list of subjects. This list includes, but is not limited to, the following:

a. Certified quality manager

b. Familiarization with company policy

c. Understanding established policy and procedures

d. Familiarization with established calibration system requirements

e. Special process training and certification

f. Applied statistical process control

g. Certified quality auditor

h. Certified quality engineer

i. Certified mechanical inspector

j. Certified quality technician

7.17.3 Location of Training

Certification and technical courses may be conducted in-house, at a college or university, or at an institution involved in continuing education programs.

7.17.4 Instructions

By a representative from internal specialists or by a qualified outside consultant.

7.17.5 Course Duration

Established by the provider of training.

7.17.6 Example of Course Descriptions

    a. Applied Statistical Process control. This course is designed to provide the student with a working knowledge regarding the application, development, interpretation, evaluation, and implementation of statistical process control techniques.

    b. Calibration System Requirement. This course provides the student with an understanding of calibration system requirements dictated by a product design and contract. Topics include the relationships between certified measurement standards, inspection, measuring and test equipment, calibration intervals, accuracy of instruments, preparation of calibration policies and procedures, and responsibilities for auditing the system.

    c. Nondestructive Inspection. This course will furnish instruction in the fundamental concepts associated with acceptance inspection of material. It presents the basic principles and theory, the applications and limitations of the various nondestructive inspection methods currently in common use. Nondestructive inspection methods, including radiography, magnetic particle, untrasonic, liquid penetrant, and electro-magnetic inspections, are discussed from a quality assurance viewpoint and applied to inspection or testing of materials.

7.17.7 Prerequisite Examples

    a. Applied Statistical Control: Enrollees in the course shall have a working knowledge of fundamental arithmetic including addition, subtraction, multiplication, and division of numbers, as well as decimal and fractional values, plus a determination of powers and roots.

b. Calibration System Requirements: Employees nominated for this course shall have at least six years in inspection and testing and have a working knowledge of arithmetic and algebraic techniques.

c. Nondestructive Testing. Personnel nominated for nondestructive testing of materials training must have completed or demonstrated an ability to read and interpret specifications and drawings, and possess at least one year of experience in manufacturing, and inspection.

7.17.8 Course Schedule

A schedule is established for all courses that are given within a fiscal year or as otherwise decided by a training monitor. It includes course titles, duration of the course in man-hours, location, starting times and the name and brief description of its contents.

7.17.9 Off-site Training

When training is provided by an independent training organization, the instructor's name, directions to the training facility, date, duration, and class location shall be provided to each trainee.

7.17.10 Record of Training

A record of training as well as future training plans is established and maintained for each employee. (See Employee Training Record, Figure 7.23.)

Note: Many local sections of the American Society for Quality (ASQ) have capable instructors that specialize in training quality managers and technicians in the disciplines of quality auditing, quality engineering, quality manager, mechanical inspector, quality technician, applied statistical process control, and many other quality related training courses.

(Continued)

Date: November 29, XXXX

Name: John McLearnit   **Home address:** Any Street, Anytown, U.S.A.

Technical skills

Discipline	Source of training
Mechanical	College and American Society for Quality (ASQ)
Electrical	College
Pneumatic	College
Hydraulic	College
Force	College
Quality Auditor	ASQ
Calibration Technician	ASQ
Quality Engineer	ASQ
Quality Manager	ASQ
Quality Engineer	ASQ

Required company training		If yes, list scheduled training data
Policy procedures	Yes ___ No _x_	_____
Calibration procedures	Yes ___ No _x_	_____
Written processes	Yes ___ No _x_	_____
Work instructions	Yes ___ No _x_	_____
Employee intern program	Yes _x_ No ___	January 29, XXXX

List of skills and years of experience: 20 years' experience in the field of quality management with the U.S. Department of Defense and private industry.

Training monitor: I.M. McTeacher

Note: Certification and registration in various disciplines of quality management should be the ultimate goal of a training program. Achieving and maintaining certifications is highly recommended.

Figure 7.23   Employee training record.

7.18 STAFF MEETINGS

Monthly meetings are held among top and middle management personnel. Topics cover appropriate areas of contract quality management. The purpose of these meetings is to support management's quest for continuous improvement of products, services, established policies, procedures, and processes and to reduce justified internal and external complaint. Management as well as nonmanagement employees are encouraged to submit agenda topics that supports their organizations quest for the improvement of established procedures and processes that are used create a products or services.

7.19 CONCLUSION

The Wee-Cando Company successfully demonstrated to the Wee-Needit Company that they have technical and administrative capability to meet contract quality requirements.

NOTES

1. ANSI/ISO/ASQ Q9004-2000: Quality management systems: Guidelines for performance improvements.
2. *Managing the Metrology System,* 2nd ed. ASQ Quality Press, page 21, 1997.

8

Self-Assessment of Managing Contract Quality Requirements Quiz

The purpose of this chapter is reinforce the information contained in this book and to support readers who are involved in day-to-day decisions associated with the administrative application of contract quality requirements. Readers who become familiar with the following Questions (Q,) answers (A), rationales (R), and references (Ref.), will be better prepared to fulfill responsibilities that relate to requests, tenders, and contracts.

Code letter	Chapter	Title
A	1	Application of a purchase plan
B	2	Communication of Contract Quality Requirements (CQR)
C	3	Quality plan for the application of CQR
D	4	Verification of contract compliance (CC)
E	5	Audit of CC
F	6	Case Study 1: Processing an invitation-for-bid
G	7	Case Study 2: Application of CQR

A. Application of a Purchase Plan

A1. Q Purchase of supplies and services are always made from suppliers who submit the lowest bid without contest.
a) True   b) False

A2. Q A comprehensive purchase plan is one that considers:
a) Buyer's needs and expectations
b) Plant manager's needs and interests
c) Seller's needs and interests
d) All of the above

A3. Q A contract administrator has sole responsibility for establishing and maintaining a purchase plan.
a) True   b) False

A4. Q When is a purchase plan prepared for a new product design?
a) Prior to the receipt of a copy of a requisition for supplies and services
b) After the receipt of a copy of a requisition for supplies and services

A5. Q When an acquisition is made for an item that is similar to one previously produced, a determination shall be made if:
a) A new procedure will be required
b) Procedures are acceptable in the as-is condition
c) Previously prepared procedures will require a change or modification
d) All of the above

A6. Q A robust purchase plan is one that:
a) Will meet a requesting activities needs and expectations
b) Includes the functions listed in a quality plan
c) Will meet the supplier's interests
d) Reflects a realistic delivery schedule
e) All of the above

A7. Q Support from a CMT is important where a purchase is complex in nature.
a) True b) False

A8. Q An all-inclusive purchase plan is one that addresses:
a) Requests
b) Tenders
c) Contracts
d) All of the above

A9. Q Which three important questions should be addressed when processing an invitation-for-bid.
a) Is there a clear description of solicited products and services?
b) Are required technical documents readily available?
c) Are required measuring devices readily available?
d) Will there be a need to conduct capability audits?
e) All of the above

A10. Q It is the buyer's responsibility to communicate what is expected from a seller.
a) True b) False

A11. Q If an audit of a proposed supplier's tender offer indicates that required resources are not readily available, the auditor will consider a tender offer to be nonresponsive to solicited requirements.
a) True b) False

A12. Q The flow-down of contract quality requirements may be distributed to management and operations personnel verbally.
a) True b) False

A13. Q A buyer's team of specialists provides support to its organization's contract administrator as to *how* each contract requirement will be met, and the seller's team of specialists advises its organization as to *what* action will be required to meet contractual obligations.
a) True b) False

A14. *Q* A customer may use three approaches toward verifying the capability of its suppliers of products and services. Name them.

   a) Suppliers who have achieved ANSI/ISO/ASQ Q9001-2000 certification status for items offered by them

   b) Suppliers certified via a first- and second-party arrangement

   c) Information provided in a potential supplier's brochure

   d) Audit of a low bidder with unknown capabilities

   e) All of the above

A15. *Q* All organizations that operate under a first- and second-party arrangement should seek ISO certification and registration.

   a) True   b) False

A16. *Q* There are three categories of suppliers. Name them.

   a) ISO certified and registered

   b) Client approved

   c) Supplier's brochure

   d) Certificate of competence

   e) All of the above

A17. *Q* A supplementary list of suppliers includes:

   a) Suppliers who are ISO certified for solicited products and services

   b) Suppliers who have a verified history within the industry for furnishing good quality on time to their customers

   c) Suppliers who have verbally requested to be added to the list

   d) Suppliers determined to be capable via a first- and second-party agreement

A18. *Q* A written purchase plan must be prepared for all products and services.

   a) True   b) False

A 19. *Q* When financial capability is an issue, there will be no need to verify this requirement in a solicitation.

   a) True   b) False

A20. *Q* A product is the only contract line item that is considered when preparing a purchase plan.

      a) True   b) False

B. Communication Between Customer and Supplier

B1. *Q* Communication of contract quality requirements begins after a first- and second-party agreement.

      a) True   b) False

B2. *Q* The pursuit of continuous improvements in the process depends on clearly communicated contract quality requirements.

      a) True   b) False

B3. *Q* Purchased items should be grouped into three general categories. Name them.

      a) Items requiring first-article approval

      b) Off-the-shelf

      c) Noncomplex

      d) Complex

      e) All of the above

B4. *Q* The primary purpose of auditing contract compliance is to assure that:

      a) Technical and other pertinent requirements are in compliance with specified requirements

      b) Stated requirements are free from omission, ambiguity, and bias

      c) Requirements of purchase plan have been met

      d) All of the above

B5. *Q* Notification of contract quality requirements to a supplier's management and operations personnel is achieved via:

      a) An abstract of contract requirements

      b) Verbal instructions

      c) A highlighted copy of a contract

      d) All of the above

B6. Q The root cause of errors of omission in the communication process may be traceable to:
   a) A purchase plan that is not established or maintained
   b) A quality plan that is not established or maintained
   c) A contract quality requirement that is overlooked during a contract review process
   d) All of the above

B7. Q Errors of omission should be reviewed and summarized by a representative of the quality control department.
   a) True   b) False

B8. Q Communication of quality policy is extended throughout the application of a quality system.
   a) True   b) False

B9. Q A review and preparation of an abstract of contract requirements and related technical documents will eliminate potential errors of omission.
   a) True   b) False

B10. Q Management planning associated with a new product design includes the following:
   a) Required sources and skills to fulfill a contractual requirement
   b) The identification of long lead items associated with a contractual obligation
   c) Need to update policies, procedures, and processes
   d) Requirement for certified status of personnel
   e) All of the above

B11. Q Errors of omission that relate to document and data control can be attributed to:
   a) Missing policy, procedures, and processes
   b) Inadequate policy, procedures, and processes
   c) Availability of technical documents
   d) All of the above

B12. *Q* A written procedure does not have to be established to identify the quality status of a product.
a) True   b) False

C. Quality Plan for the Application of Contract Quality Requirements

C1. *Q* Suppliers of products and services prepare maintain a quality plan that describes how its organization complies with a product specification.
a) True   b) False

C2. *Q* Suppliers of products and services prepare and maintain a quality plan that describes how an organization will verify compliance with a contractual obligation.
a) True   b) False

C3. *Q* When significant changes occur in a product or other contract requirements a planner will:
a) Reexamine the plan
b) Make appropriate changes to the plan
c) Do nothing

C4. *Q* Quality objectives are achieved via a robust quality plan.
a) True   b) False

C5. *Q* Special processes subject to certification requirements do not have to be included in a quality plan.
a) True   b) False

C6. *Q* If the acquisition of a new item(s) will necessitate the need for additional plant facilities and/or equipment, they may be developed by a supplier or they may be acquired from an outside source.
a) True   b) False

C7. Q Plans for the application of a calibration system include:
a) Required measuring instruments
b) Responsibility for maintaining instrument accuracy
c) Tightest product tolerance
d) Required instrument accuracy
e) All of the above

C8. Q Human resources allocation may have to be considered when a purchase is made for new products that that are complex in nature.
a) True   b) False

C9. Q A quality plan does not have to be modified when supplementary contract quality requirements are subsequently introduced in a procurement document.
a) True   b) False

C10. Q A design plan is reviewed periodically to provide a current model of a design configuration.
a) True   b) False

C11. Q Design reviews are conducted by technical specialists who are familiar with the essential elements of a design requirement.
a) True   b) False

C12. Q First-article planning is intended to minimize the risks of producing nonconforming products and services during the production of a item.
a) True   b) False

C13. Q First-article planning is also accomplished to assure that manufacturing processes employed, workmanship standard utilized, and methods employed will meet the expectations of a plant manager.
a) True   b) False

C14. Q A documented report of first-article observations addresses:
   a) Technical documents
   b) Contract/purchase order number
   c) Test results
   d) All of the above

C15. Q A quality plan should identify a procedure for the control of documents as associated with:
   a) Product design
   b) Administrative operations
   c) Operation functions
   d) All of the above

C16. Q Technical documents do not have to be approved when management and operations personnel are familiar with their assignments.
   a) True   b) False

C17. Q Organizations solicited to design and supply a new product will adopt a model for quality assurance that address:
   a) Design, development, production, installation, and servicing
   b) Production, installation, and servicing
   c) None of the above

C18. Q Procedures for planning and implementing quality audits are implemented to determine the effectiveness of a quality system.
   a) True   b) False

C19. Q When an auditor finds a contract line item to be nonconforming, the product is identified and returned to the producer of the item.
   a) True   b) False

C20. Q A quality plan is prepared prior to the receipt of a contact award for products and services.
   a) True   b) False

C21. *Q* The purpose of establishing a contract review procedure is to coordinate contract quality requirements with a third party.
a) True   b) False

C22. *Q* Evidence that a planned process is achieving desired results should be supported with documented evidence.
a) True   b) False

C23. *Q* Delegation of certain contract requirements to a subcontractor might include:
a) All instrument calibration requirements
b) A portion of instrument calibration
c) Packaging and shipping
d) All of the above

C24. *Q* Measuring instruments that are calibrated by an independent calibration laboratory need not be a component of an organizations quality plan.
a) True   b) False

C25. *Q* Processes should be carried out under controlled conditions.
a) True   b) False

C26. *Q* Planning for a research and development program requires a full-blown application of a quality system.
a) True   b) False

C27. *Q* Plans for the control of customer-supplied products might include a product as well as inspection, measuring, and test equipment.
a) True   b) False

C28. *Q* Quality assurance orientation conferences are held between buyer and seller to:
a) Review a tender offer
b) Clarify a contract requirement
c) Review a statement of work
d) All of the above

C29. Q Quality assurance conferences are directed toward assuring that nothing in a quality plan is left open to assumption or interpretation.

a) True b) False

C30. Q The person responsible for notifying the need for a quality assurance conference is also responsible for correcting open items resulting from the conference.

a) True b) False

C31. Q The primary purpose of convening a quality assurance orientation conference is to determine the need for contract change notices.

a) True b) False

D. Verification of Contract Compliance

D1. Q Verification of contract compliance starts after the completion of:

a) Written procedures
b) Purchase plan
c) Processes
d) Quality plan

D2. Q There are three up-front actions that must be considered when verifying contract compliance. Verification of capability, the need for a post-award conference and a mutual agreement between buyer and seller.

a) True b) False

D3. Q Verification of contract compliance includes the assessment of data generated via inspection and testing.

a) True b) False

D4. Q Name three reasons for generating inspection and testing data.

a) Keeps the client happy
b) Used for measuring process capability
c) It is one of the focal points for evaluating the investigation of customer complaints
d) Used to measure process variability

D5. *Q* A certificate of conformance associated with the delivery of raw material stating only that the material meets specifications is considered acceptable.

   a) True   b) False

D6. *Q* A customer-supplied product is one that is:

   a) A product

   b) Special tooling

   c) Special inspection, measuring, and test equipment

   d) All of the above

D7. *Q* When a contract references a requirement for a special process, procedures shall be established to ensure that certified processes, test methods, equipment, and personnel meet special qualifications.

   a) True   b) False

D8. *Q* When it is determined that a certification of a special process is required, a prime contractor will have its own personnel and equipment certified.

   a) True   b) False

D9. *Q* When concurrent inspection and/or testing between buyer and seller is a contractual obligation, the time and date of this action is determined by the buyer.

   a) True   b) False

D10. *Q* Calibration system requirements are implemented at:

   a) Prime contractor's plant

   b) Subcontractor's facility

   c) Independent calibration laboratory

   d) All of the above

D11. *Q* Producers of services should have a comprehensive plan for the application of all segments of a calibration system that is based on:
a) Contract quality requirements
b) Clients interests
c) Required measuring instruments
d) Product parameters
e) All of the above

D12. *Q* Recommendations for the adjustment of calibration intervals are made by:
a) Calibration laboratory
b) Subcontractor
c) User of an instrument
d) All of the above

D13. *Q* Procedures used to calibrate inspection, measuring, and test equipment are outlined in an organization's policy manual.
a) True   b) False

D14. *Q* When it is determined that a calibration procedure is not supplied by the manufacturer of an instrument, it will be prepared by:
a) Prime contractor
b) Independent calibration laboratory
c) Independent testing laboratory
d) All of the above

D15. *Q* Product characteristics, general purpose instruments, and primary standards must be addressed when preparing a procedure associated with the traceability of measuring devices.
a) True   b) False

D16. Q   Handling, storage, packaging, preservation, and delivery parameters should be documented in the same manner as the recording of product characteristics.
a) True   b) False

E. Audit of Contract Quality Requirements

E1. Q   The basic elements of a quality system that an auditor looks for at the outset of an audit includes quality policy, procedures, processes, and management support.
a) True   b) False

E2 Q   Knowledge of the requirements imposed in a contract is essential when selecting pertinent audit factors.
a) True   b) False

E3. Q   Audit checklists associated with policy, procedures, and processes are prepared during the establishment of a quality plan.
a) True   b) False

E4. Q   A customer audits its quality policy after the selection of a potential supplier of products and services.
a) True   b) False

E5. Q   Second- and third-level audits should be:
a) Objective
b) Meaningful
c) Related to a contractual obligation
d) All of the above

E6. Q   Audits checklists are subject to modification.
a) True   b) False

E7. Q Quality audits are designed to:
- a) Examine the ability of a process to continually produce conforming products and services that conform to contract quality requirements
- b) Identify an organization's strength and weaknesses
- c) Identify opportunities for improvements
- d) All of the above

E8. Q Audits are conducted by a team of quality assurance specialists.
- a) True
- b) False

E9. Q Audit of products and services that are noncomplex are treated in the same way as complex/critical items.
- a) True
- b) False

E10. Q Audit checklists are prepared at product verification stations.
- a) True
- b) False

E11. Q In addition to identifying areas of compliance within a quality system, what else is identified in a audit report?
- a) Areas of noncompliance
- b) Observations and conclusions
- c) How to rectify nonconformity
- d) All of the above

E12. Q Audit frequency is determined by the owner of a process.
- a) True
- b) False

E13. Q A supplier's policy, procedures, and processes are audited before an operation and related activity is put into effect.
- a) True
- b) False

E14. *Q* In the event that a required procedure has been found to be inadequate or not available to operations personnel when needed:

a) The person who retains the authority to approve the procedure will be immediately notified regarding observes condition(s)

b) It will have a negative impact on the delivery of an end item

c) It adds unanticipated costs to the bottom line

d) All of the above

E15. *Q* Recommended corrective action should relate to open contracts.

a) True   b) False

E16. *Q* On-the-spot action to correct a defect shall be accomplished only when:

a) A defect is considered minor

b) The quality assurance representative considers that follow-up action in not necessary

c) None of the above

d) a and b

E17. *Q* A breakdown in a quality plan requires assurance that the quality of the products and services is not compromised.

a) True   b) False

E18. *Q* An acceptable report in response to a corrective action request is one that:

a) Responds within a recommended time frame

b) Identifies and corrects reported defects and deficiencies

c) Eliminates causes

d) All of the above

E19. *Q* Audits are designed to examine the ability of established policy, procedures, and processes to:

a) Consistently produce conforming services

b) Consistently produce conforming products

c) Identify opportunities for improvements

d) All of the above

F. Case Study 1: Processing an Invitation-For-Bid

F1. *Q* A customer will assess the capabilities of its suppliers of products and services after the award of a contract.
a) True   b) False

F2. *Q* A tender offer that requires an audit of its capabilities takes priority over routine audits.
a) True   b) False

F3. *Q* The full cycle time allowed for processing a tender offer where capabilities are unknown is:
a) At the discretion of the auditor
b) Usually seven days
c) At the discretion of the buyer

F4. *Q* There may be situations where capabilities of suppliers who submit a tender offer may request a capability assessment after a contract is awarded.
a) True   b) False

F5. *Q* Decisions made regarding a proposed supplier's capability are based on:
a) Competence
b) Capability
c) Responsibility
d) All of the above

F6. *Q* When auditing a low bidder's capability, the auditor may discuss his/her observations with a proposed supplier.
a) True   b) False

G. Case Study 2: Application of Contract Requirements

G1. *Q* Contract requirements are always summarized on an abstract of copy of a contract.
a) True   b) False

G2. *Q* Quality planning is accomplished prior to the receipt of a contract and a statement of work.
a) True   b) False

G3. *Q* Instructions for conducting inspection and testing include:
a) Item nomenclature
b) Part number
c) Product characteristic
d) Product characteristic tolerance
e) Measuring device
f) All of the above

G4. *Q* A certificate of compliance received from a supplier is considered acceptable if it includes a general statement that raw material is in compliance with specifications and that the material meets the physical and chemical analysis requirement.
a) True   b) False

G5. *Q* Suppliers are listed in two general categories. Name them.
a) Those suppliers who produce complex items
b) Those suppliers who produce complex items and services
c) Those suppliers who produce noncomplex items
d) Those suppliers who produce noncomplex items and services

G6. *Q* The accountability of inspection and acceptance stamps is the responsibility of a plant manager.
a) True   b) False

G7. Q A document control procedure is established and maintained to assure that the latest drawings, specifications, and contract change notices are available when needed.
a) True   b) False

G8. Q Audit functions are centered on the needs and interests of the management functions of a supplier and not the customer.
a) True   b) False

G9. Q Organizations conduct periodic staff meetings to:
a) Support management's quest for continuous improvements in its quality system
b) Review justified internal and external complaints
c) Assure customer satisfaction
d) All of the above

G10. Q Agenda topics submitted by non-management personnel should be encouraged.
a) True   b) False

G11. Q Engineering is solely responsible for assuring that services furnished to clients do in fact conform to a purchase agreement.
a) True   b) False

G12. Q The sole purpose of convening among key managers to achieve continuous improvement of the services provided to clients.
a) True   b) False

ANSWERS – RATIONALE – REFERENCES

Chapter 1: Application of a Purchase Plan
Code Letter A

A1. False

R. When purchasing supplies and services from an apparent low bidder with questionable capabilities, a potential customer will finalize a purchase only after verifying that the low bidder can meet solicited contract quality requirements.

Ref. Chapter 1, Clause 1.1

A2. a and c

R. The acquisition of supplies and services should be coordinated with a plan that not only considers the seller's needs and interests but also meets a buyer's needs and expectations at a reasonable cost.

Ref. Chapter 1, Clause 1.2

A3. False

R. A contract administrator, or where appropriate a purchasing manager, assumes the responsibility for implementing and maintaining a purchase plan with support from a contract quality management team.

Ref. Chapter 1, Clause 1.2

A4. b

R. A purchase plan is prepared after the timely receipt of a requisition for solicited products and services. When an acquisition is made for an item that is similar to one previously produced, the planner will assure that existing procedures do or do not require or modifications.

Ref. Chapter 1, Clause 1.1.1

A5. d

R. When an acquisition is made for a previously produced product or service, the purchaser with support from appropriate members of a CQMT determine if existing procedures will or will not require a change or modification.

Ref. Chapter 1, Clause 1.12

A6. a and c

R. A robust purchase plan is one that not only meets a requesting activities needs and expectations but also meets a supplier's interests.

Ref. Chapter 1, Clause 1.2

A7. True

R. Support is particularly important where solicited and services are complex.

Ref. Chapter 1, Clause 1.2

A8. d

R. A strong purchase plan is one that includes procedures for the review of requests (an invitation-for-bid), tenders (a response to an invitation-for-bid), and contracts (contract administration).

Ref. Chapter 1, Clause 1.12.1

A9. a, b, and d

R. Procedures associated with preparing an invitation-for-bid include, but are not limited to, the following:

a) Description of solicited products and services

b) Required technical documents

c) Method of packaging and shipping

d) Product/service inspection and acceptance points

e) Post-award orientation conference

f) Identification of supplementary contract quality requirements

Ref. Chapter 1, Clause 1.13

A10. True

R. Buyer's responsibility begins with the receipt of an all-inclusive purchase requisition and the acquisition of supplies and services from capable suppliers.

Ref. Chapter 1, Clause 1.13

A11. False

R. When resources are not available at the time of a pre-contract audit, a proposed supplier can meet solicited requirements by acquiring them from an outside source. However, delivery of the required resources must be provided within the time constraints of a solicitation and backed up with written quotes that are considered to be acceptable by the auditor.

Ref. Chapter 1, Clause 1.14.1

A12. False

R. Contract quality requirements should be reviewed, documented, and distributed in a timely manner. Distribution is accomplished after it is determined that contract-related technical requirements are clearly defined.

Ref. Chapter 1, Clause 1.16

A13. False

R. Members of a buyer's contract quality management team provide assistance to its organization's contract administrator as to what will be expected from a supplier. Conversely, a seller's team of specialists provides support as to how each contract requirement will be met.

Ref. Chapter 1, Clause 1.18

A14. a, b, and d

R. A potential customer wants to look at a supplier's applicable administrative and manufacturing capabilities and how its resources will be applied to meet specified requirements.

Ref. Chapter 1, Clause 1.20

A15. False

R. Not necessarily. Some buyers and sellers who are content to operate under a first- and second-party arrangement for the following reasons:

a) A purchaser is dealing with a supplier who has established an exceptionally good quality record within the industry

b) A purchaser is doing business with a supplier who is the owner of a proprietary item and has a history of delivering good quality

c) A prime contractor's quest for continuous improvement techniques requires direct contact with his/her suppliers

Ref. Chapter 1, Clause 1.22.1

A16. a, b, and d

R. Suppliers are listed as follows:

1) Suppliers who have achieved American National Standards Institute (ANSI), the International Standards Organization (ISO), and the American Society for Quality (ASQ) certified and registered status

2) Suppliers who are not ISO-certified but have previously provided sufficient evidence of their capabilities to their clients in advance of a contract negotiation

3) Reliance on a certificate of conformance

Ref. Chapter 1, Clause 1.23

A17. a, b, and d

R. A supplementary list of suppliers identifies an organization's anticipated need for new suppliers who are capable of meeting a stated requirement for a product or service. The list identifies organizations that have submitted a written request to be included in a potential client's list of capable suppliers. The selection process is based on the following three conditions:

1) Supplier is ISO certified for products or products and services offered

2) Supplier has a history within the industry of supplying quality products and services

3) An audit of suppliers who are not ISO certified is required prior to adding the organization to the client's list of approved suppliers

Ref. Chapter 1, Clause 1.24

A18. False

R. A plan is prepared for the purchase of complex items. However, a written plan may not be required for noncomplex/noncritical items. There are two categories of purchases that must be considered when preparing a purchase plan:
1. Development and maintenance of a written purchase plan for purchasing supplies and services that are complex
2. Functions for the purchase of noncomplex items that do not require a written purchase plan

Ref. Chapter 1, Clause 1.10

A19. False

R. Where there is a need to verify financial capability, it should be included in a solicitation for products and services.

Ref. Chapter 1, Clause 1.5.1

A20. False

R. When preparing a new purchase plan for a new product or service, the planner should take into account items previously produced that are similar in nature.

Ref. Chapter 1, Clause 1.11

Chapter 2: Communication Between Customer and Supplier
Code Letter B

B1. False

R. Communication begins well in advance of customer/supplier interaction. It starts with top management's commitment to quality excellence and with a competent work force that respects and develops everyone's capabilities.

Ref. Chapter 2, Clause 2.2

B2. True

R.   The pursuit of continuous improvements in the process to create a product depends on clearly communicated contract requirements between the activity requesting a buy and the buyer, between the buyer and the seller, between the seller and its team of contract administrators, and between the seller and its subcontractors.

Ref. Chapter 2, Clause 2.2

B3. b, c, and d

R.   The purpose of grouping purchases is to identify those contract line items that will require special attention— particularly items with characteristics that are complex.

Ref. Chapter 2, Clause 2.4

B4. d

R.   A customer's contract administrator, with support from a team of specialists and an auditor, is responsible for verifying that an established purchase plan will be met.

Ref. Chapter 2, Clause 2.5

B5. a and c

R.   A supplier establishes and maintains an effective contract- review procedure that is developed in conjunction with all other management functions, then provides a summarized version of a contract to key managers.

Ref. Chapter 2, Clause 2.6

B6. d

R.   Errors of omission can (and usually do) lead to nonconforming processes, poor product quality, and unnecessary repair and rework of products and services.

Ref. Chapter 2, Clause 2.7.3

B7. True

    R. Errors of omission and customer complaints should be monitored. This action is taken to identify trends that adversely impact established procedures and processes and to identify and correct the cause.

    Ref. Chapter 2, Clause 2.9

B8. True

    R. The communication process should address appropriate elements of an adopted quality system standard and associated applicable factors of a quality system.

    Ref. Chapter 2, Clause 2.11

B9. True

    R. Contracts, technical documents, and established policy, procedures and processes that are free of errors and omissions will eliminate stumbling blocks that originate from contractual and related documentation that are suspect.

    Ref. Chapter 2, Clause 2.7.1

B10. e

    R. When management fails to establish and maintain an organization of capable specialists or fails to provide key managers with specific responsibility, authority, and empowerment, it can lead to systemic problems and customer complaints.

    Ref. Chapter 2, Clause 2.7.9

B11. d

    R. Errors of omission can be attributed to missing or inadequately written procedures that define the process for issuing and retrieving administrative and technical documents. They also prevail when required documents are not readily available at a respective product/service verification station.

    Ref. Chapter 2, Clause 2.7.12

B12. False

> **R.** Procedures are established for the satisfactory identification of the quality status (good or bad) of a product. When procedures are ignored, a supplier runs the risk of comingling good products with the bad ones.
>
> **Ref.** Chapter 2, Clause 2. 7.14

Chapter 3: Quality Plan for the Application of Contract Quality Requirements
Code Letter C

C1. False

> **R.** Suppliers of products and services prepare and maintain a quality plan that describes how its organization will ensure compliance with a contractual obligation.
>
> **Ref.** Chapter 3, Clause 3.2

C2. True

> **R.** A quality plan serves as a master quality plan and a control document.
>
> **Ref.** Chapter 3, Clause 3.2

C3. a and b

> **R.** A planner will review a plan that was previously prepared for a similar item to determine if the information contained therein might be incorporated in the new plan.
>
> **Ref.** Chapter 3, Clause 3.2

C4. True

> **R.** Quality objectives and financial gains are the results of the application of a robust quality plan that is blended with documented policies and procedures.

C5. False

R. Quality plans identify where and how special processes, such as heat treating, magnetic particle inspection, welding, and so on, will be inspected, tested, and controlled.

Ref. Chapter 3, Clause 3.2

C6. True

R. When new plant facilities and/or equipment will be needed to produce a contract line item, a quality plan will indicate if they will be developed within the supplier's facility or of if they will be acquired from an outside source within the time constraints of a negotiated contract.

Ref. Chapter 3, Clause 3.4

C7. e

R. The plan identifies the adopted calibration system standard, required general purpose instruments, and required reference and transfer standards, as well as the tightest product tolerance.

Ref. Chapter 3, Clause 3.5

C8. True

R. When a contract dictates the need for an organization to hire additional employees, a planner, with the help of appropriate members of a contract quality management team, should determine the number and type of employees that will be needed. Needs might include skilled, unskilled and administrative personnel. The availability and source of human resources should also be identified in the plan.

Ref. Chapter 3, Clause 3.6

C9. False

R. Modification of an original product design is an example of supplementary contract quality requirements. When a modification to a purchase is issued, its impact on a completed quality plan must be determined.

Ref. Chapter 3, Clause 3. 7

C10. True

    R. Engineering evaluates the effects of design changes and techniques utilized to improve the intrinsic reliability of a design.

    Ref. Chapter 3, Clause 3.9.1

C11. True

    R. The nature and/or complexity of a proposed design will dictate what specialists will be present during a design review.

    Ref. Chapter 3, Clause 3.9.4

C12. False

    R. First-article review of production and pilot models is accomplished to assure that a supplier can furnish a product that is satisfactory for its intended use and to reduce the risks of producing nonconforming products and services.

    Ref. Chapter 3, Clause 3.10

C13. False

    R. First-article review is accomplished to determine whether a producer of an item clearly interpreted specified requirements and is able to furnish a product that is satisfactory for its intended use.

    Ref. Chapter 3, Clause 3.10

C14. d

    R. Organizations who perform first-article inspection and testing are obligated to prepare a report of each inspection and test that shows technical documents that were used to check the first article. In addition, the report addresses other information such as the applicable contract or purchase order, where inspections and tests were performed, as well as results in qualitative terms.

    Ref. Chapter 3, Clause 3.10

C15. d

R. A document change-control procedure is maintained for the control of documents associated with a product design, as well as administrative and operation functions.

Ref. Chapter 3, Clause 3.11

C16. False

R. An all-inclusive quality plan will include technical documents described in sufficient detail so that management and operations personnel can perform their assigned functions in a timely manner. Documents should be reviewed for clarity and completeness, and approved by a designated representative of an organization before they are put into use.

Ref. Chapter 3, Clause 3.12

C17. a

R. The selection of an appropriate quality system model is predicated on the design issues. If a supplier is required to design and supply a product and service then he or she will select a quality system standard that covers all of the quality elements that relate to design development. b) When a supplier is solicited to furnish a product to an established design, he or she will establish procedures and processes that relate to quality elements associated with production, installation, and servicing

Ref. Chapter 3, Clause 3.14

C18. True

R. Quality audits are established and maintained to verify whether quality activities and related results comply with planned arrangements and to determine the effectiveness of a quality system.

Ref. Chapter 3, Clause 3.15

C19. False

R. When a product is found to be nonconforming, it is identified as such and held for review action by a preliminary review monitor. Where appropriate, the nonconforming product or lot will be processed through a material review board.

Ref. Chapter 3, Clause 3.16

C20. False

R. Quality plans are prepared after the receipt of a signed contract and related technical documents. The plan may be developed on a plant-wide scale, and it should incorporate sufficient flexibility to accommodate individual contract requirements.

Ref. Chapter 3, Clause 3.18

C21. False

R. The purpose of establishing a contract review procedure is to coordinate contract quality requirements, second-party management, and operations personnel.

Ref. Chapter 3, Clause 3.18

C22. True

R. Records associated with the application of processes include, but is not limited to, quality policy, procedures, purchase, production and quality plans, audit reports, control charts, inspection and test results, conference agendas, and related reports.

Ref. Chapter 3, Clause 3.18

C23. d

R. A delegation may include a part or all of a contractual requirement and is determined by the prime contractor (second party).

Ref. Chapter 3, Clause 3.23

C24. False

    R. Instruments that are calibrated in-house as well as those that are calibrated by an independent calibration laboratory are identified is a supplier's quality plan for each new product design or contract line item.

    Ref. Chapter 3, Clause 3.26

C25. True

    R. Processes are established and maintained to assure that the production of products and services will be carried out under controlled conditions and to verify that they are being produced in accordance with specified requirements.

    Ref. Chapter 3, Clause 3.3

C26. False

    R. A quality plan is a progressive plan that evolves as a product and service is developed.

    Ref. Chapter 3, Clause .3.9.1

C27. True

    R. Customer-supplied items include both products and measuring instruments.

    Ref. Chapter 3, Clause 3.2.4

C28. b and c

    R. Quality assurance orientation conferences and held when:

        a) Information in a contract in not clear enough to make a determination regarding responsibility or application.

        b) A requirement exceeds the latest state-of-the-art standards.

        c) Tailored contract requirements that will require input from a buyer's technical specialist.

    Ref. Chapter 3, Clause 3.31.1

C29. True

> **R.**   A quality assurance conference aids key personnel in achieving a clear and mutual understanding of a statement of work and related contract quality requirements.
>
> **Ref.** Chapter 3, Clause 3.31

C30. False

> **R.**   The individual who determines that a quality assurance conference (QAC) will be required identifies the time and place of a conference, prepares an agenda, and notifies participants. The owners of a process are responsible for rectifying open items.
>
> **Ref.** Chapter 3, Clause 3.31.5

C31. False

> **R.**   It is not the intent of a QAC to convene for the sole purpose of requesting a contract change notice. Its purpose is to identify and clarify contractual requirements.
>
> **Ref.** Chapter 3, Clause 3.33

Chapter 4: Verification of Contract Compliance

Code Letter D

D1. a, c, and d

> **R.**   Verification of contract compliance starts after the completion of policy, procedures, processes, and a quality plan.
>
> **Ref.** Chapter 4, Clause 4.4.2

D2. True

> **R.**   Verification of supplier capability is required when (1) a supplier does not have ANSI/ISO/ASQ Q9001-2000 certified status, (2) a post-award orientation will be needed, where necessary, to assure that contract quality requirements are clearly understood, and (3) a mutual agreement exists between buyer and seller regarding the interpretation, each contract requirement.
>
> **Ref.** Chapter 4, Clause 4.4.3

D3. True

> **R.** Inspection and test-related data are used in concert with the application of procedures and processes which in turn support the assessment of pertinent quality elements associated the verification of contract compliance.
>
> **Ref.** Chapter 4, Clause 4.5.1

D4. b, c, and d

> **R.** There are several reasons for conducting product inspection and testing. It is for measuring process capability, process variability, and audits.
>
> **Ref.** Chapter 4, Clause 4.5.1

D5. False

> **R.** Where a certificate of conformance is a requirement of a contract or purchase order, a receiving inspector will accept the material only when it is determined that a certification is accompanied with test data and physical and chemical analysis reports, and that they are found to meet a specified requirement.
>
> **Ref.** Chapter 4, Clause 4.8.3

D6. d

> **R.** A customer-supplied product includes assemblies, components, raw materials, measuring equipment, and supplies that may be consumed during contract compliance.
>
> **Ref.** Chapter 4, Clause 4.8.4

D7. True

> **R.** A requirement for the application of a special process is established and maintained as a result of a requirement referenced in drawings, specifications, or other documents developed by a supplier or as specified in a contractual agreement.
>
> **Ref.** Chapter 4, Clause 4.9.5

D8. False

R. A supplier might elect to have its own personnel certified or he or she may elect to delegate the responsibility to an independent testing laboratory.

Ref. Chapter 4, Clause 4.9.6

D9. False

R. Notification of the time and date that a product is ready for concurrent source inspection or testing is made via a written notice from the seller.

Ref. Chapter 4, Clause 4.10.2

D10. d

R. Calibration requirements are implemented at a prime contractor's facility or as delegated to a subcontractor's or an independent calibration laboratory.

Ref. Chapter 4, Clause 4.11.1

D11. a, c, and d

R. Planning defines how contract quality requirements will be met within an overall quality system that addresses product characteristics and their code numbers; the measuring devices used to inspect a product characteristic; and the instruments code number.

Ref. Chapter 4, Clause 4.11.1

D12. c

R. Interval adjustment is based on usage. Therefore, the final decision regarding the establishment and subsequent adjustment of a calibration interval remains solely with the user of instruments.

Ref. Chapter 4, Clause 4.11.3

D13. False

R. Procedures used for the calibration of measuring instruments are usually prescribed in an instrument manufacturer's instruction manual.

Ref. Chapter 4, Clause 4.11.2

D14. d

    R.   Calibration procedures that are not readily available from an instrument manufacturer will be prepared by the producer of a product or an independent testing and calibration laboratory.

    Ref. Chapter 4, Clause 4.11.2

D15. True

    R.   Traceability of measuring instruments include:

        a) Product characteristics to general purpose instruments (GPI)

        b) GPI to transfer standards (TS)

        c) TS to primary standards (PS)

        d) PS to the National Institute of Standards and Technology

    Ref. Chapter 4, Clause 4.11.1

D16. True

    R.   A checklist should be used by an inspector when verifying that a product is ready for shipment. Selected inspection checklists must relate to a contract's quality requirement.

    Ref. Chapter 4, Clause 4.11.1

Chapter 5: Audit of Contract Quality Requirements

Code Letter E

E1. True

    R.   Audits are intended to assure both internal and external customers that established policy, procedures and processes are achieving defined objectives.

    Ref. Chapter 5, Clause 5.1.1

E2. True

    R.   A robust audit is one that centers its focus on the evaluation of product parameters and systemic factors that are most representative of a contractual requirement.

    Ref. Chapter 5, Clause 5.1.4

E3. False

R.   The basis for conducting a quality audit are the checklists that are prepared during desk audit planning.

Ref. Chapter 5, Clause 5.2.1

E4. False

R.   Customers audit their own policies and procedures prior to selecting suppliers of products and services so as to assure that their needs and expectations are not compromised.

Ref. Chapter 5, Clause 5.2.2

E5. d

R.   An organization's needs should be objective and meaningful, and its checklists should be directly related to specific contract requirements.

Ref. Chapter 5, Clause 5.3

E6. True

R.   Audit checklists may be modified or improved subsequent to feedback recommendations made by the owner of a process, or from justified customer complaints and nonconforming reports generated by the quality control department.

Ref. Chapter 5, Clause 5.3.1

E7. d

R.   Opportunities for improvements are gained when quality assurance specialists who conduct an audit have a comprehensive knowledge of the procedures and processes associated with contract specified design, development, production, installation, and servicing of products and services.

Ref. Chapter 5, Clause 5.5.1

E8. False

R.   Audits may be performed by an individual or a team of individuals who are familiar with contract quality requirements; applicable quality system standards; and related policy, procedures, and processes.

Ref. Chapter 5, Clause 5.6

E9. False

    R. The degree of audit input is predicated on the complexity of a product design as well as the complexity of a process. Hence, contract-related products and services that are complex in nature, require more time and manpower for implementation.

    Ref. Chapter 5, Clause 5.7

E10. False

    R. Audit planning begins by during the selection of appropriate checklists during a desk audit.

    Ref. Chapter 5, Clause 5.12

E11. a and b

    R. In addition to identifying areas of compliance within a quality system, an audit report also identifies areas of noncompliance, conclusions, and recommendations. Corrections are the responsibility of the planner and the owners of a process.

    Ref. Chapter 5, Clause 5.13

E12. False

    R. Audit frequency is based on the complexity of a product design; criticality of applicable product characteristics; and the needs, interests, and expectations of both customer and supplier.

    Ref. Chapter 5, Clause 5.17.1

E13. True

    R. Documentation is reviewed prior to the application of an operation to determine if a procedures or process will or not fulfill the requirements stated in a contractual agreement. Each contract should be reviewed to determine contract quality requirements and required action.

    Ref. Chapter 5, Clause 5.17.2

E14. d

> **R.** Objective management requires that quality procedures are prepared in a timely manner, that they are all-inclusive and that they are readily available when needed.

> **Ref.** Chapter 5, Clause 5.17.2

E15. True

> **R.** Recommendations should be factual, objective, and related to active contract quality requirements.

> **Ref.** chapter five, clause 5.17.3

E16. a and b

> **R.** The extent of corrective action is dependent on the seriousness and importance of a reported defect and its impact on form, fit, or function.

> **Ref.** Chapter 5, Clause 5.17.3

E17. True

> **R.** Deficiencies, other than those that are minor in nature, should be documented and reported to applicable department leaders for necessary action. The report should show the date that the deficiency was detected, a reference number for identification purposes, description of the deficiency, the impact on open contracts, the effect on delivery schedules, and recommended response date.

> **Ref.** Chapter 5, Clause 5.17.3

E18. d

> **R.** A director of quality assurance or a designated representative assumes the responsibility for verifying that satisfactory corrective action is taken to rectify a reported complaint.

> **Ref.** Chapter 5, Clause 5.17.3

E19. d

> **R.** Audits are designed to examine the ability of established procedures and processes to consistently produce conforming products and services.

> **Ref.** Chapter 5, Clause 5.51

Chapter 6: Processing an Invitation-For-Bid

Code Letter F

F1. False

    R. Prior to issuing a contract to a supplier with unknown quality assurance, financial, or production capabilities, a potential customer should audit the potential supplier before issuing a signed contract.

    Ref. Chapter 6, Clause 6.3.1

F2. True

    R. Turnaround time for assessing the capabilities of a tender offer is very short; action must be taken within the time specified in a solicitation.

    Ref. Chapter 6, Clause 6.3.3

F3. b

    R. Empirical knowledge indicates that the time allowed for processing a tender offer is usually seven working days. The process starts with the performance of an audit and ends with conclusions made by the buyer's purchasing manager and/or a standing review board.

    Ref. Chapter 6, Clause 6.3.3

F4. False

    R. The normal process is for a buyer to audit capability where it is unknown. When this requirement is a component of a solicitation, any exceptions to a tender offer can be considered as nonresponsive.

    Ref. Chapter 6, Clause 6.3.4

F5. d

    R. Conclusions and recommendations made in an audit report must be free from ambiguity and differences of opinions and must be based on a proposed supplier's current and past experience that relate to solicited items.

    Ref. Chapter 6, Clause 6.3.5

F6. False

> **R.** The director of purchasing or a designated representative will notify the bidder after audit conclusions and recommendations are made by a standing review board.
>
> **Ref.** Chapter 6, Clause 6.9

Chapter 7: Application of Contract Requirements
Code Letter G

G1. False

> **R.** Contract requirements are usually summarized on an abstract of contract requirements. However, a highlighted copy of a purchase order is used for summarizing noncomplex products and services.
>
> **Ref.** Chapter 7, Clause 7.3.1

G2. False

> **R.** Planning is accomplished after it is found that contract requirements are clearly understood and that associated technical documents are readily available to management and operations personnel.
>
> **Ref.** Chapter 7, Clause 7.4

G3. f

> **R.** Written inspection and test procedures should be made available at each product verification station. Instructions include item nomenclature, part number, product characteristic, characteristic tolerance, and related measuring instruments.
>
> **Ref.** Chapter 7, Clause 7.5.1

G4. False

> **R.** Raw material is accepted only when it is accompanied with a detailed certificate of compliance report.
>
> **Ref.** Chapter 7, Clause 7.5.3

G5. b and d

   R.   Suppliers are listed into two general categories: those that produce complex items and services, and those that produce noncomplex items and services. The reason for this list is to provide quality assurance personnel and the director of purchasing with advance notice regarding the need for special planning associated with items that are complex in nature.

   Ref. Chapter 7, Clause 7.6.3.1

G6. False

   R.   The issue and retrieval of inspection stamps that are used to indicate inspection status is the responsibility of a quality control manager.

   Ref. Chapter 7, Clause 7.10

G7. True

   R.   An engineering department assumes the responsibility for releasing and controlling all parts lists, drawings, specifications, bills of material, and operation sheets as well as the procedure and process documents used to manufacture parts and assemblies.

   Ref. Chapter 7, Clause 7.11

G8. False

   R.   It is recognized that the bulk of quality audits are focused on the quality management functions of the supplier. However, customer audits are not ignored, particularly when there is a need to verify capability.

   Ref. Chapter 7, Clause 7.12

G9. d

   R.   Regularly scheduled meetings are convened to support management's quest for continuous improvements of established policies and procedures, to consider beneficial suggestions, and to determine how customer complaints can be rectified.

   Ref. Chapter 7, Clause 7.18

G10. True

R.   Management and non-management employees are encouraged to submit agenda topics that support an organization's quest for continuous improvements in all of the areas of contract quality management.

Ref. Chapter 7, Clause 7.18

G11. False

R.   Engineering and production quality control departments, along with a customer complaint monitor, are delegated with the responsibility for assuring that services furnished to a clients are continually evaluated.

Ref. Chapter 7, Clause 7.16

G12. False

R.   Meetings are held among top and middle management personnel. The purpose of these meetings is to continuously improve services, as well as product, policy, procedures, and processes.

Ref. Chapter 7, Clause 7.18

Glossary

accuracy—The closeness of agreement between an observed value and an accepted value.[1]

audit—Systematic, independent, and documented progress for obtaining audit evidence and evaluating it objectively to determine the extent to which the audit criteria are fulfilled.

auditee—The organization being audited.[2]

audit team—One or more auditors conducting an audit.

auditor—Person with competence to conduct an audit.

calibration—The set of operations that establish, under specific conditions, the relationship between valued indicated by a measuring instrument or measuring system, or values represented by a material measure or a reference material, and the corresponding values of a quantity realized by a reference standard.[3]

capability—Ability of an organization, system, or process to realize a product that will fulfill the requirements for that product.

characteristic—A property that helps to differentiate between items of a given sample population. *Note:* The differentiation may be either quantitative (by variables) or qualitative (by attributes).[1]

comparator—An instrument for comparing some measurement with a fixed standard.[5]

conformity—Fulfillment of a special requirement.[2]

contractor—Supplier in a contractual situation.[2]

contract review—Systematic activities carried out by a supplier before signing the contract to ensure that requirements for quality are adequately defined, free from ambiguity, documented, and can be realized by the supplier.[2]

corrective action—Action taken to eliminate the causes of an existing nonconformity, defect, or other undesirable situation in order to prevent recurrence.[2]

customer—Recipient of a product provided by a supplier.[2]

defect—Non-fulfillment of an intended usage requirement of reasonable expectations, including one concerned with safety.[2]

degree of documentation—Extent to which evidence is produced to provide confidence that qualified requirements are fulfilled.[2]

disposition of nonconformity—Action taken to deal with an existing nonconforming entity in order to resolve the nonconformity.[2]

document—Information and its medium.

element—A quality of a product, material, or service forming a cohesive entity on which a measurement or observation may be made.

error of measurement—The result of a measurement minus the value of the measurement.[3]

hold point—Point defined in an appropriate document, beyond which an activity must not proceed without the approval of a designated organization authority.[2]

inspection—The process of measuring, examining, testing, gaging, or otherwise comparing the unit with the applicable requirements.[1]

inspection by attributes—Inspection by attributes in inspection whereby either the unit of product is classified as conforming or nonconforming, or the number of non-conformities in the product is counted, with respect to a given requirement or a set of requirements.[4]

international standard—A standard recognized by an international agreement, that serves as the basis for fixing the value of all other standards to the quality concerned.[3]

item—An object or quantity of material on which a set of observations can be made, or the results of making an observation of an object or quantity of material.[1]

limits of permissible error (of a measuring instrument)—The extreme values of an error permitted by a specification, regulation, and so on, for a given measurement.[3]

management—Coordinated activities to direct and control an organization.

measurement control system—Set of interrelated or interacting elements necessary to achieve metrological confirmation and continued control of measurement processes.

measurement process—Set of operations to determine the value of a quantity.

measurement standards (MS)—A material measure, measuring instrument, reference material, or system intended to define, conserve, or reproduce a unit or more values of a quantity in order to transmit them to other measuring instruments by comparison.[3]

measuring equipment—All of the measuring instruments, measurement standards, reference materials, auxiliary apparatus, and instructions that are necessary to carry out a measurement. This includes measuring equipment used in calibrations.[3]

metrological characteristic—Distinguishing feature that can influence the results of a measurement.

metrological confirmation—Set of operations required to ensure that a measuring instrument is in a state of compliance with the intended requirements.[3]

metrology—The science of measurement.[5]

model for quality assurance—Standardized or selected set of quality system requirements.[2]

national standard—A standard, recognized by national agreement, that serves as a basis for fixing the value of all other standards to the quality concerned.[3]

nonconformity—Nonfulfillment of a specified requirement.[2]

organizational structure—Responsibilities, authorities, and relationships arranged in a pattern through which an organization performs its functions.[2]

preventive action—Action taken to eliminate the causes of a potential nonconformity, defect, or other undesirable situation in order to prevent recurrence.[2]

procedure—Specified way to carry out an activity or a process.

process—Set of interrelated or interacting activities which transforms inputs to outputs.

purchaser—Customer in a contractual situation.[2]

qualification process—Process of demonstrating whether an entity is capable of fulfilling specified requirements.[2]

qualified—Status given to an entity when capability of fulfilling specified requirements has been demonstrated.[2]

quality—The totality of features and characteristics of a product or service that bears on its ability to satisfy given needs.[1]

quality assurance—A systematic and independent examination to determine whether quality activities and related results comply with planned arrangements, and whether these arrangements are implemented effectively and are suitable to achieve objectives.[3]

quality auditor—Person qualified to perform audits.[2]

quality management—The totality of functions involved in the determination and achievement of quality.[1]

quality plan—Document setting out the specific quality practices, resources, and sequence of activities relevant to a particular product or contract.[2]

quality policy—Overall intentions and direction of an organization with regard to quality, as formally expresses by top management.[2]

quality related costs—Those costs incurred in ensuring satisfactory quality, as well as the losses incurred when satisfactory quality is not achieved.[2]

quality system—Organizational structure, procedures, processes, and resources needed to implement quality management.[2]

reference material—A material or substance, of one or more properties, which is sufficiently well established to be used for the calibration of apparatus or a measurement method or for the assignment of values to materials.[3]

specification—Document-setting requirements.[2]

subcontractor—Organization that provides a product to the supplier.[2]

total quality management—Management approach by an organization that is centered on quality, based on the participation of its members, aimed at long-term success through customer satisfaction, and beneficial to all members of the organization and to society.[2]

traceability—Ability to trace the history, application, or location of an entity by means of recorded identification.[2]

variables, method of—Measurement of quality by measuring and recording the numerical magnitude of a quality characteristic, for each of the units in the group under consideration. This involves reference to a continuous scale of some kind.[1]

verification—Confirmation by examination and provision of objective evidence that specified requirements have been fulfilled.[2]

SOURCE OF GLOSSARY TERMS

ASQ Statistical Division, *Glossary and Tables for Statistical Quality Control* 3d ed., Milwaukee: ASQ Quality Press, 1996.

1 ANSI/ISO/ASQC A8402-1994, *Quality management and quality assurance – Vocabulary.*

2 ISO 10012-1:1992: *Quality assurance requirements for measuring equipment – Part 1: Metrological confirmation system for measuring equipment.*

3 ANSI/ASQC Z1.4, *Sampling Procedures and Tables for Inspection by Attributes*

4 *Webster's New Dictionary of the American Language.* Springfield MA: G&C Merriam Co., 1975.

5 ANSI/ISO/ASQ Q9000-2000, *Quality management systems – Fundamentals and vocabulary.*

Acronyms

ANSI	American National Standards Institute
ASQ	American Society for Quality
CCM	Customer complaint monitor
CMT	Contract management team
CQR	Contract quality requirement
CSP	Customer-supplied product
DoD	Department of Defense
FA	First article
FAR	Federal acquisition regulation
FMEA	Failure mode effective analysis
FOB	Free on board
GPI	General purpose instrument
GSA	General Service Administration
IS	International Standard
M&TE	Measuring and test equipment
MD	Measuring device
MRB	Material review board
MRL	Master requirements list
NCSL	National Conference of Standards Laboratories
NIST	National Institute of Standards and Technology
NT	Negative trend
PR	Preliminary review
PS	Primary standard
PSI	Pounds per square inch
QAC	Quality assurance conference
QPL	Qualified product list
SAS	Seek another source
SOP	Standard operation procedure
SPC	Statistical process control
TS	Transfer standard

Bibliography

ANSI/ASQ Z1.4: Sampling Procedures and Tables for Inspection by Attributes, 1993.

ANSI/ASQ Z540-1: American National Standard for Calibration: Calibration Laboratories and Measuring and Test Equipment: General Requirements.

ANSI/ASQC E2-1996: Guide to Inspection Planning.

ANSI/ISO 17025-1999: Calibration requirements for the competence of testing and calibration laboratories.

ANSI/ISO/ASQ 10011: Guidelines for auditing quality systems, 1994.

ANSI/ISO/ASQ Q9001-2000: Quality management systems: Requirements.

ANSI/ISO/ASQ Q9004: Quality management systems: Guidelines for improvements.

ASQ Statistics Division. *Glossary and Tables for Statistical Quality Control.* 4th ed., 2004.

BS EN ISO 14971-2001: Medical devices: Application of risk management to medical devices.

Campanella, Jack. *Principles of Quality Costs: Principles, Implementation, and Use,* 2nd Ed., Milwaukee: ASQ Quality Press, 1999.

Defense DoD-Std-480A: Configuration Control: Engineering Changes, Deviations, and Waivers, 1978.

Department of Defense, Defense Supply Agency. DSAM 8200.1 Procurement Quality Assurance, August 1976.

Department of Defense, Evaluation of a Contractor's Calibration System Handbook MIL-HDBK 52, August 1989.

Department of Defense. General Services Administration, and National Aeronautics and Space Administration. Government Federal Regulation, March, 1995.

Department of Defense. Handbook H50: Evaluation of a Contractor's Quality Program, 1965.

Department of Defense. Mil-Std 120 Gage Inspection, September, 1963.

Department of Defense. Mil Std 45662A, Calibration System Requirements, August, 1988.

International Organization for Standardization. ISO 8402 International Standard.

ISO 9000:2000: Quality management systems: Fundamentals and vocabulary.

ISO 10012-1:1992: Quality assurance requirements for measuring equipment: Metrological confirmation system for measuring equipment.

ISO 10012-2:1997: Quality assurance for measuring equipment – Part 2: Guidelines for control of measurement processes.

ISO 10012:2003: Measurement management systems: Requirements for measurement processes and measuring equipment.

ISO 13485:2003 Medical devices – Quality systems – Requirements for the regulatory purposes.

ISO/TS 16949: Quality management systems: Particular requirements for the application of ISO Q9001-2000 for automotive production and relevant service part organizations.

Quality Management and Quality Assurance Vocabulary, 1994.

C. Robert Pennella, *Managing the Metrology System*, Milwaukee, WI: ASQ Quality Press, 3rd ed, 2004.

Software Quality Assurance Program Requirements MIL-S-52779A, August 1, 1979.

Index